AGRICULTURE ISSUES

MANURE: MANAGEMENT, USES AND ENVIRONMENTAL IMPACTS

AGRICULTURE ISSUES AND POLICIES

Series Editor: Carmen S. Dellaguardia

Additional books in this series can be found on Nova's website at:

https://www.novapublishers.com/catalog/index.php?cPath=23_29&seriesp=
Agriculture+Issues+and+Policies

Additional E-book in this series can be found on Nova's website at:

https://www.novapublishers.com/catalog/index.php?cPath=23_29&seriespe=
Agriculture+Issues+and+Policies

AGRICULTURE ISSUES AND POLICIES

MANURE: MANAGEMENT, USES AND ENVIRONMENTAL IMPACTS

CARMEN S. DELLAGUARDIA
EDITOR

Nova Science Publishers, Inc.
New York

Library of Congress Cataloging-in-Publication Data
Manure : management, uses and environmental impacts / editor: Carmen S. Dellaguardia.
p. cm.
Includes index.
ISBN 978-1-61668-424-2 (hardcover)
1. Farm manure--Management. 2. Farm manure--Environmental aspects. 3. Manure handling. I. Dellaguardia, Carmen S.
S655.M35 2010
631.8'61--dc22

2010004374

Published by Nova Science Publishers, Inc. ✝ New York

CONTENTS

Preface		**vii**
Chapter 1	Manure Management: Issues and Options *Edward Osei*	**1**
Chapter 2	Manure: Its Impacts on Environment Quality and on Soil Biota with Special Reference to Nematodes *Qudsia Tahseen*	**49**
Chapter 3	Phosphorus Forms in Animal Manure and the Impact on Soil P Status *Zhongqi He and Zhengxia Dou*	**83**
Chapter 4	Improving Performance and Manure Management in the French Pig Sector: A Three-Stage Analysis *Isabelle Piot-Lepetit*	**115**
Chapter 5	Similar Bacterial Community Structure and High Abundance of Sulfonamide resistance Genes in Field-Scale Manures *Chu Thi Thanh Binh, Holger Heuer, Newton Carlos Marcial Gomes, Martin Kaupenjohann and Kornelia Smalla*	**141**

Chapter 6 Bioremediational Technique of Adding Nitrite-Oxidizing Bacteria for Reducing N_2O Emission during Swine Manure Composting **167**
Yasuyuki Fukumoto

Index **181**

PREFACE

Manure is organic matter used as organic fertilizer in agriculture. Manure contributes to the fertility of the soil by adding organic matter and nutrients, such as nitrogen that is trapped by bacteria in the soil. As the interest in organic farming and renewable energy production continues to grow, manure is playing an important role in the production of bioenergy, food, and other agricultural products. This new book discusses varied topics on the management, uses and environmental impacts of manure, such as: the bioremediational technique of adding nitrite-oxidizing bacteria for reducing N2O emission during swine manure composting; improving performance and manure management in the french pig sector; manure management in Africa; phosphorous forms in animal manure and the impact on soil P status; and others.

Chapter 1- Manure is an important resource for crop production. Manure includes feces, urine, and bedding from livestock and other animals (animal manure), residue or biomass from plants (plant manure), and decomposed forms of either animal or plant manure (compost). While all three main types of manure play an important role in agriculture, animal manure has received a much greater attention due to its nutrient content, and its potential impact on the environment and on human health. This chapter presents a comprehensive review of the issues and options related to manure management, with particular emphasis on manure generated on livestock production operations.

As the interest in organic farming and renewable energy production continues to grow, manure is playing an increasing role in the production of bioenergy, food, and other agricultural products. However, the role of manure as crop nutrient source, soil amendment, feedstock for renewable energy, bedding, livestock feed additive, and other useful purposes, is tempered by its potential impact – particularly livestock manure – on water and air quality in many watersheds that have a significant presence of livestock operations. To minimize the potential adverse effects of manure, many management options have been designed and implemented on livestock operations, particularly in developed nations. However, the effectiveness of these options is predicated upon proper maintenance over time, of the practices implemented. Cost considerations and complex biophysical factors often hinder the effectiveness of manure management options. Furthermore, while interactions among certain practices may have synergistic effects, other combinations of management options often exert contradictory effects on each other.

Effective policy design that takes into account the biophysical context of farms and the long-term economic implications of alternative management practices show greater promise for improved use of manure resources. In particular, policies that are holistic in nature, those that offer flexibility in the choice of manure handling technologies, and those that promote market-based programs that integrate farmer accountability for manure management are preferred to more restrictive and costly programs. To support cost-efficient management of manure in developed countries, emerging economies, and developing nations alike, more data and user-friendly tools are needed to rapidly and reliably determine the costs and ecosystem impacts of alternative manure management options.

Chapter 2 - Soil health can be boosted by improving soil texture, aeration, drainage, and nutritional content by application of manure/amendments besides promoting the growth of the beneficial biota and suppressing the pathogens. Manures can vary in their nutrient contents and forms, depending on their processing and storage procedures, feeds and feed supplements used, etc. Of the total N, 30% to 90% is comprised of ammonium while the rest is contained in organic forms that must be decomposed (mineralized) to inorganic forms to be available to plants. Likewise, the phosphorus in manure tends to be fixed in soils by sorption and precipitation reactions and its mineralization takes place to make it available to plants. Manures are also effective sources of potassium (K) and sulphur (S). The manure-enriched soils enhance the populations of mites and collembolans besides increasing the number of fungi that directly affect the population of soil nematodes. Some of the fungi use adhesive conidia, branches, knobs and mycelia to parasitize nematodes. However, compared to plant parasitic nematodes, the density of opportunistic bacterial-feeding and fungal-feeding nematodes increases significantly in the soils treated with manure. Among the manures of animal origin poultry and farmyard manures give significantly higher yields than those of other organic manures. The impacts of manures also vary considerably in the soil systems. Poultry manure causes short term (within a growing season) increases in microbiological activity and nutrient cycling, whereas farm yard manure seems to bring long-term changes. Green manure application too has gained wide acceptance due to the improvement in soil characteristics, including drainage and water retention, nutrient content and storage, and level of microbiological activity in the soil.

Chapter 3 - Increased knowledge of manure phosphorus (P) chemistry is needed to optimize nutrient recycling and to minimize adverse environmental effects associated with land application of manure. Part I of this chapter reviews and discusses various manure P characterization methods including sequential fractionation, solution ^{31}P nuclear magnetic resonance (NMR) spectroscopy, enzymatic hydrolysis, and solid state spectroscopic techniques. We then present a case study that compares P forms in conventional and organic dairy manure identified by solution and solid state ^{31}P NMR spectroscopy. The results of this case study suggest that the lability of P may differ between organic and conventional dairy manures; therefore specific management practices may be required for organic dairy manure. Part II of the chapter focuses on the impacts of manure application on soil P status. For this part we present and discuss the results of laboratory aerobic incubation and field studies to evaluate short- and long-term impacts, respectively. The lack of stable organic P in relevant extracts of the laboratory or field soils with manure amendment indicates possible degradation of stable manure P. Sequential fractionation of pasture soils reveals that the levels of labile P are more closely related to the number of years receiving poultry litter, whereas the levels of stable P are related to the cumulative amount of poultry litter applied. The two different patterns imply that the impact of manure application on the environment (including runoff potential) should be gradually factored in with increasing length of application.

Chapter 4 - Nitrogen emissions from agriculture are considered as an important problem, motivating an evaluation of the performance of French pig farms in improving manure management. This article develops a methodology that allows for a joint evaluation of farms' economic and environmental efficiency from various productive structures. A three-step approach is developed. First, a Cluster Analysis is implemented. It provides a description of main French pig farm structures. Farms are grouped into clusters with similar structural characteristics regarding land, labour and production diversification versus specialisation. Second, a Data Envelopment Analysis is implemented at the level of each cluster. This nonparametric frontier analysis allows for efficiency benchmarking. It provides a score on the economic performance of farms and on their efficiency in manure management in a way that reduces the detrimental impact of their productive activity on the environment. At last, an aggregate procedure is developed to measure the performance at the cluster level and at the sector level. This bottom-up approach allows us to provide an overall indicator of French pig farms' performance in both an economic and environmental direction and to identify the percentage of this performance that depends on farms' characteristics or on the structure of the pig sector.

Main results show that improvements both in economic and environmental terms are limited for farming systems specialized in pig production without any change in their production structure or technology of production. Farming systems that are diversified can contribute to an increase in production and a decrease in nitrogen surplus by implementing their current production technology in a most efficient way.

Chapter 5- The bacterial community structure and antibiotic resistance of bacteria in 16 spread manures originating from 15 pig producing facilities in Germany were investigated. The farms varied in numbers of pigs, production type, and antibiotic usage. Between 2 to 69% of cultured bacteria were resistant to the recalcitrant sulfonamide antibiotic sulfadiazine, while less than 0.3% showed resistance to the much less stable β-lactam antibiotic amoxicillin. The abundance of the β-lactam resistance gene *bla*-TEM, and the sulfonamide resistance genes *sul1* and *sul2* in total community DNA was quantified by real time PCR relative to 16S rRNA gene copies. In all manures, a high proportion of bacteria carried *sul* genes, especially *sul2*, while the relative abundance of *bla*-TEM was at least two log units lower.

Bacterial community fingerprinting by PCR-DGGE of 16S rRNA gene fragments revealed that the dominant bacterial populations were surprisingly similar in all the manure samples. In contrast, enterobacterial patterns varied much more between samples, indicating that this group of potential pathogens is more susceptible to factors differing among piggery manures. Cloning and sequencing of 16S rRNA gene fragments reamplified from dominant DGGE bands, and of 16S rRNA genes amplified from total community DNA showed that the dominant populations in manure belonged to the Clostridia. In conclusion, substantial amounts of antibiotic resistant bacteria are released into the environment by spread manure, which contributes to the cycling of resistance genes between animal production facilities, human communities and clinical environments, resulting in an increasing threat to antibiotic therapy.

Chapter 6- Livestock waste treatment holds a large nitrous oxide (N_2O) potential with global emission. Composting is a principal treatment method of solid fraction of swine manure in Japan. However, swine manure composting entails substantial emissions of nitrogenous compounds including N_2O. During swine manure composting, N_2O is mainly emitted after thermophilic phase because the nitrification mediated by mesophilic microorganisms (ammonia-oxidizing bacteria [AOB] and nitrite-oxidizing bacteria [NOB]) is a necessary prerequisite for N_2O generation from fresh manure which contains little nitrate (NO_3^-) nitrogen. However, the growth of indigenous NOB after thermophilic phase tends to be late than that of AOB, which could be a cause of toxic nitrite (NO_2^-) accumulation during maturation phase of composting. It is known that there is a strong correlation between N_2O emission and NO_2^- accumulation in the composting process; therefore, the NO_2^- accumulation after thermophilic phase induces substantial N_2O emission. For the purpose of shortening the time lag of proliferation between AOB and NOB, a bioremediational technique of adding mature compost, including NOB, at the start of AOB proliferation was examined in the experiments of swine manure composting. The adding mature compost after thermophilic phase made it possible to establish NOB guild more quickly in the composting material, which could prevent NO_2^- accumulation by prompt oxidization from NO_2^- to NO_3^-. As the pattern of N_2O emission agreed well with that of NO_2^- concentration, N_2O emission ceased rapidly when NOB was added. Summarizing data of laboratory-scale experiments, the average reducing rate of N_2O emission by the mature compost (NOB source) addition from the control was calculated to be 60%, and therefore, this technique appeared to allow a quantitative reduction in N_2O emission in the swine manure composing. From the result of nitrogen mass balance in the composting experiment, it was found that the NO_2^- accumulation induced not only N_2O, but also other nitrogenous emissions during compost maturation; therefore, a possibility that if NO_2^- accumulation has been prolonged the nitrogen loss induced by NO_2^- accumulation could become a principal cause of the total nitrogen losses was suggested. To reduce such a serious environmental risk as global warming and improve the efficiency of nitrogen from animal compost, the bioremediational technique for complete nitrification promotion is considered to be effective for mitigation of N_2O emission from swine manure composting.

In: Manure: Management, Uses and…
Editors: C. S. Dellaguardia, pp. 1-48 ISBN: 978-1-61668-424-2

Chapter 1

MANURE MANAGEMENT: ISSUES AND OPTIONS

Edward Osei
Tarleton State University

ABSTRACT

Manure is an important resource for crop production. Manure includes feces, urine, and bedding from livestock and other animals (animal manure), residue or biomass from plants (plant manure), and decomposed forms of either animal or plant manure (compost). While all three main types of manure play an important role in agriculture, animal manure has received a much greater attention due to its nutrient content, and its potential impact on the environment and on human health. This chapter presents a comprehensive review of the issues and options related to manure management, with particular emphasis on manure generated on livestock production operations.

As the interest in organic farming and renewable energy production continues to grow, manure is playing an increasing role in the production of bioenergy, food, and other agricultural products. However, the role of manure as crop nutrient source, soil amendment, feedstock for renewable energy, bedding, livestock feed additive, and other useful purposes, is tempered by its potential impact – particularly livestock manure – on water and air quality in many watersheds that have a significant presence of livestock operations. To minimize the potential adverse effects of manure, many management options have been designed and implemented on livestock operations, particularly in developed nations. However, the effectiveness of these options is predicated upon proper maintenance over time, of the practices implemented. Cost considerations and complex biophysical factors often hinder the effectiveness of manure management options. Furthermore, while interactions among certain practices may have synergistic effects, other combinations of management options often exert contradictory effects on each other.

Effective policy design that takes into account the biophysical context of farms and the long-term economic implications of alternative management practices show greater promise for improved use of manure resources. In particular, policies that are holistic in nature, those that offer flexibility in the choice of manure handling technologies, and those that promote market-based programs that integrate farmer accountability for manure management are preferred to more restrictive and costly programs. To support cost-efficient management of manure in developed countries, emerging economies, and developing nations alike, more data and user-friendly tools are needed to rapidly and reliably determine the costs and ecosystem impacts of alternative manure management options.

INTRODUCTION

Manure is any material that fertilizes land; especially refuse of stables and barnyards consisting of livestock excreta with or without litter (Webster, 2009). Broadly defined, manure includes feces, urine, and bedding from livestock and other animals (*animal manure*), residue or biomass from plants (*plant manure*), as well as decomposed forms of either animal or plant manure (*compost*). For thousands of years manure has been an integral part of crop production around the world. Before the discovery of synthetic fertilizers the vast majority of crop nutrients were supplied by manure sources and legume nitrogen fixation. While synthetic materials now account for a greater share of plant nutrients in use today, manure is still a significant plant nutrient source in every nation.

Manure is a more holistic source of plant nutrients than synthetic fertilizers. In addition to nitrogen, phosphorus, and potassium, manure contains many micronutrients and other beneficial substances. In addition, manure provides a more conducive medium for the growth of beneficial microorganisms than synthetic fertilizers. It also supplies organic matter to soils and can help reduce soil erosion. Furthermore, notwithstanding the well-accepted definitions for manure, an increasing role for manure in recent years is its use as feedstock in bioenergy production, which has been fueled by a demand for renewable energy sources.

In spite of their numerous benefits, manures present a number of problems to the livestock or crop farmers who produce the manure as well as to those who use it. Particularly in the case of livestock operations, animal manure is increasingly being produced on operations that do not have adequate land for its beneficial use, and this often results in significant environmental pressures on these farms. Animal manure also harbors various pathogens, and may contain weed seeds and other harmful substances that can serve as a liability to users of the manure. Furthermore, due to the low concentrations of nitrogen, phosphorus, and potassium in manure as compared to synthetic fertilizers, manure can be quite costly to transport and use in today's intensive agricultural systems.

In recent years, due to an increased interest in organic farming and organic farm produce, the range of uses for manure has broadened. Manure and various manure products that result from varying degrees of processing, are now used not only as plant nutrient source, but also in organic materials that are advertised as having pesticidal properties, as mulch, in various nursery products, and in packaging material. Manure is also used in the livestock and poultry feed industry as a useful ingredient in animal feeds. Concomitant with the increased scope of manure use, manure products are no longer largely confined to small sized operations. Manure products are now integrated in commercial plant feed ingredient and animal feed production processes that supply feed additives to large livestock operations.

The contrasting benefits and potential liabilities of manure production and use call for appropriate management of this significant plant nutrient resource in order to maximize the benefits derived from it and minimize any associated drawbacks. This chapter presents the issues and options associated with manure management. While crop and livestock sources of manure are both significant, greater focus will be given here to animal manure. The benefits, issues, and options related to manure are presented to help the reader to gain a better understanding of how best to harness manure resources for beneficial use. While numerous references are cited throughout the chapter, this is not intended to be a comprehensive review of the manure management literature.

Beneficial Uses of Manure

Manure has many benefits. Besides its obvious use as crop nutrient source, animal and crop (green) manures can be used in a wide variety of ways. As the interest in organic farming and ecosystem health continues to rise, an increasing spectrum of manure benefits is becoming more apparent. Unfortunately, for most people, manure is often associated with negative connotations – principally bad odor and water and air pollution nuisances. Thus the benefits of manure, however varied and broad in scope, are not obvious to many. To readers who are unfamiliar with the benefits of manure use, it is hoped that this section will provide a better perspective of the potential benefits of this biological resource.

Table 1. Selected manure production characteristics (lb/AU/day) for major livestock and poultry species[1]

Characteristic	Dairy	Beef	Swine	Broiler	Turkey	Sheep
Total manure	86	58	84	85	47	40
Total solids	12	8.5	11	22	12	11
Volatile solids	10	7.2	8.5	17	9.1	9.2
Total Kjeldahl Nitrogen	0.45	0.34	0.52	1.1	0.62	0.42
Total phosphorus	0.094	0.092	0.18	0.3	0.23	0.087
Potassium	0.29	0.21	0.29	0.4	0.24	0.32
Calcium	0.16	0.14	0.33	0.41	0.63	0.28
Magnesium	0.071	0.049	0.07	0.15	0.073	0.072
Sulfur	0.051	0.045	0.076	0.085	***	0.055
Chloride	0.13	***	0.26	***	***	0.089
Iron	0.012	0.0078	0.016	***	0.075	0.0081
Manganese	0.0019	0.0012	0.0019	***	0.0024	0.0014
Boron	0.00071	0.00088	0.0031	***	***	0.00061
Zinc	0.0081	0.0011	0.0050	0.0036	0.015	0.0016
Copper	0.00045	0.00031	0.0012	0.00098	0.00071	0.00022

[1] Based on ASABE (2005)

While this section provides a comprehensive list, it is not meant to be an exhaustive discussion of the beneficial uses of manure. The benefits are listed here in more or less decreasing order of relevance with respect to the proportion of manure put to various uses. The emphasis here is on animal manures, though green manures also play a significant role in many areas. Other publications that provide useful reviews of the beneficial uses of manure include USDA-ARS (2009) and EPA, 2000??).

Current Uses of Manure

Crop nutrient

Perhaps the best known role of manure is its use as a crop nutrient source. Unlike synthetic fertilizers, manure contains many micronutrients and other beneficial substances and organisms in addition to nitrogen, phosphorus, and potassium. The American Society of Agricultural and Biological Engineers tabulates typical fresh manure characteristics in their Engineering Practices Standards publication (Table 1). Manure nutrients are also released gradually over time, making them less prone to rapid leaching and runoff losses (particularly nitrogen) than synthetic fertilizer sources of the same nutrients.

Renewable energy source

For many centuries animal manure was used on a small scale to generate energy primarily for farm use. In recent years, due in part to the increase in demand for and interest in renewable energy sources, the number of on-farm bio-digestion systems has grown significantly. Many of the recently developed farm-based bio-energy generation systems have been installed on larger scaled operations due to the significant capital investments entailed (Osei and Tanter, 2007). Alternative technologies that can be used to convert manure to useable bio-energy include anaerobic digestion (USEPA, 1997; Lusk, 1998), gasification (USEPA, 2009; Clark, 2006), cofiring, and methanol (USEPA, 2009).

Soil amendment/conditioner

In contrast to synthetic fertilizers, manure possesses significant soil amendment properties. When applied to the soil, manure adds organic matter. A study in Ontario showed that application of dairy manure at 20 tons/acre will increase soil organic matter levels by about 0.065% per year, based on a

6-inch soil sample depth (Hayes, 2003). The organic matter added to the soil through manure application enhances the soil texture and other soil properties of both clay and sandy soils. Manure increases the water holding capacity of sandy soils, improves water infiltration in clayey soils, and also reduces evaporation in drought conditions. Manure also encourages root development and reduces soil erosion by improving formation of soil aggregates. Soil cation exchange capacity (CEC), pH levels, and pore space all improve with addition of appropriate amounts of manure.

Manure amends the soil also because it contains numerous natural substances and organisms that are favorable to the growth of beneficial soil organisms. Various microorganisms and bacteria that are crucial to the long-term health of the soil ecosystem thrive more favorably in manured soils than those receiving synthetic fertilizers. Manure itself often contains significant populations of these beneficial organisms.

Animal feed additive

A number of commercial livestock operations use processed manure, particularly processed poultry litter, in their livestock feed rations. When added to livestock feed, processed poultry litter is a significant protein source. In this use poultry litter is known to have a much higher value than its use as a crop nutrient.

Mulch

Composted animal manure can also be used as mulch in fruit and vegetable production. For this purpose the manure needs to be fully cured (thoroughly composted) to avoid introduction of weed seeds into the garden or farm. The composted manure can be mixed with other ingredients to increase its effectiveness as mulch.

Livestock bedding

After some processing, manure can be used as part of livestock bedding material. For this purpose composted or otherwise cured manure is used in order to avoid re-introduction of harmful pathogens that reside in fresh manure back into the livestock housing system.

Algae production

There is growing interest in the use of liquid effluent from lagoons on dairy farms to grow algae for various uses. Algae produced from liquid manure effluent can be processed into livestock feed additives or crop fertilizer materials. Another prominent use of algae grown on liquid waste is the generation of biodiesel. Several efforts are ongoing in the United States and other countries to commercialize algae production for the multiple purposes including biodiesel production, feed and fertilizer production, and extraction of excess nutrients from the liquid waste stream.

Carbon sequestration

When manure is left on the soil surface, it contributes to greenhouse gas emissions by emitting methane and various nitrogenous compounds, chiefly nitrous oxides). Emission of methane from manure piles adds to the global carbon footprint. On the contrary, applying manure for beneficial use in crop production will increase soil organic matter contents, thus increasing the amount of carbon sequestered in the soil and reducing the carbon footprint.

Nursery materials

Manure has been used as potting material in the manufacture of nursery pots. While only of somewhat limited use, this is a rather natural use of manure, as it renders the pot largely biodegradable and the nutrients and substances released are actually beneficial to the soil medium receiving the transplants.

Aquaculture

In a number of Asian countries, manure has been used in aquaculture for many years. Manure serves as a useful source of nutrient for the biological communities in aquatic environments. Use of manure in aquaculture enhances the growth of the flora and fauna in aquatic environments and consequently leads to improved populations of the fish that thrive on those plant and animal organisms.

Mushroom cultivation

USEPA (2009) also lists mushroom cultivation as one of the uses of composted manure. CAST (1996) reports that the use of composted manure in commercial mushroom production has been proven to be safe; however, economic factors limit its usefulness. Sutton et al. (1999) indicate that horse manure is most often used in this application.

Soil reclamation

Studies have shown that manure can be used successfully for soil reclamation (USEPA, 2009). An experimental study with burnt soil showed that application of poultry manure was instrumental in reclamation of the soil, as compared to the unmanured control (Vázquez, 1996). Green manures have also been used in successful soil reclamation efforts.

Sealing ponds and dams

USEPA (2009) reports that manure has also been used to seal ponds and dams in a number of countries, including Russia. After the manure has been applied over the inner base and walls of the dam, the manure layer is covered with other organic materials, and allowed to sit undisturbed for several weeks, during which period anaerobic bacteria act on the manure pile. The dam is ready to be flooded after the anaerobic process has been completed.

Building materials

Researchers led by Deland Myers at Iowa State University have succeeded in developing fiber board from cow manure (Barron, 2000). While the process is still experimental, there appears to be remarkable success in transforming manure from cows into useful renewable composites for a variety of uses.

Innovative figurines from composted manure

Hannawald (1999) reports that a pouiltry producer in Arkansas discovered an innovative way of molding composted poultry litter into figurines of ducks, turtles, swans, and cats. The producer sells the figurines to customers who place them in their yards. As the ornaments disintegrate over time, the manure nutrients and useful soil amendments are released into the soil. In this case, a farmer who had difficulty finding adequate land for litter disposal turned that problem into a success story through an appropriate innovation.

Paper production

Manure has also been used in the manufacture of paper products. Though, often on a small scale, horse and sheep manure has been used in the manufacture of paper in various countries including Great Britain and the United States. Manure used for these processes is often from animals that consume lower concentrations of protein in their diet. For odor reasons manure from animals such as elephants that excrete lower concentrations of nitrogen is more desirable for paper production. Certain greeting cards have been known to be manufactured from manure from such animals.

Other beneficial uses of manure

With the increased interest in organic farming and gardening in recent years, numerous web sites and small scale commercial outlets have advertised various manure products for the treatment of plant diseases and the control of pest damage to crops. Though controlled scientific experiments are yet to confirm the validity of most of these claims (Chalker-Scott, 2007), anecdotal evidence from a number of consumers suggests that some of these claims are probably true. Among the manure products that are advertized for plant disease and pest control, compost teas are one of the most prominent. One notable research trial indicating success with compost teas is an experiment on organic pecan production in Comanche County, Texas, by USDA Agricultural Research Service (USDA-ARS) scientists that was reported in the Pecan South Magazine in 2008 (Flores, 2008). In this study, the use of compost teas was indicated to improve plant health and control insects and diseases of the pecan crop (Flores, 2008).

Compost teas are produced by the aerobic or anaerobic fermentation of manure for a period of time that can range from several hours to a few weeks. Typically, some manure is put in a container, which is then topped with water and allowed to sit for a period of time (Garrett, 2009). With aerobic fermentation the resulting liquid is stirred occasionally. After the manure has undergone this process of fermentation, the liquid is poured through a strainer into another container. The strained liquid is then diluted with additional water and sprayer on the plants to be treated. Some compost teas are fortified with additional ingredients such as garlic to enhance their effectiveness for the specific purpose intended. Advocates of compost teas indicate that it works because of the natural and holistic communities of beneficial substances and microorganisms that are contained in the teas.

Potential for Manure Use in the U.S.

Using current data on livestock and poultry inventories, a reliable estimate can be obtained of the potential crop nutrients from manure sources. Similarly, the potential for bio-energy generation from animal manure sources can be estimated from current data on livestock numbers.

Potential crop nutrient from manure sources

Aggregate manure nutrients available for crop nutrient use can be estimated by applying reasonable loss estimates to the manure nutrient production values (Table 1) presented earlier. Manure nutrients available for crop utilization are simply obtained as the product of manure produced by each livestock species and the specific nutrient losses associated with the type of manure management practice being used on the farm. Specifically, for each farm, and for each nutrient *N*,

$$N_a = N_f \prod_{k=1}^{s} (100 - N_k^i)/100^s$$

where N_a is the total amount of nutrient *N* available for plant utilization,

N_f is the amount of nutrient *N* in fresh manure,

N_k^i is the impact of the kth management practice on nutrient *N* (percentage losses), for each of the *s* management options employed on the farm, and

$\prod$ is the product summation symbol.

Once typical storage and handling losses have been applied for various nutrients, studies indicate that manure remains a significant source of nitrogen, phosphorus, and potassium for crop production in the U.S. Based on 1997 data used in PPI (2002), it is estimated that in the absence of storage and handling nutrient losses, manure could satisfy 40% of crop nitrogen needs – 12.9 billion pounds of nitrogen, 77% of crop phosphate requirements (8.8 billion pounds), and 54% of crop potash uptake (10.4 billion pounds) on an annual basis. Based on 2008 average commercial fertilizer nutrient prices, manure nutrients potentially available for plant uptake are valued around $20 billion per year in the U.S. alone. It should be noted that this figure does not account for the ancillary benefits of manure use other than crop nutrient uptake, such as its value as a soil amendment as indicated above.

Potential bio-energy generation from manure sources

Collectible manure generated on confined livestock and poultry operations can be used to generate electrical energy or heat. The recent years have seen an increased interest in on-farm and centralized digesters that convert methane to electrical energy. While only a very small portion of U.S. farms have operating digesters, data on livestock inventories indicate that there is great potential to generate billions of kilowatt hours (kWh) of electricity each year using animal manure.

Table 2. Potential energy generation from U.S. livestock and poultry[1]

Animal type	Animal units	Biogas energy per animal unit/day[2]		Biogas energy/year[2]
	(Millions)	(1000 BTU)	(kWh)	(Billion kWh)
Fattened Cattle	9.6	8.5	2.5	8.71
Milk cows	12.3	6.8	2.0	8.94
Other beef and dairy cattle	58.8	7.7	2.2	48.14
Swine	8.5	13.1	3.8	11.94
Poultry	6.1	18.5	5.4	12.06
Total	95.3	54.5	16.0	89.79

[1]Based on data reported in Cuéllar and Webber (2009) and Kellogg et al. (2000).
[2]Assumes 33% biogas to energy conversion efficiency.

Table 3. Beneficial uses of manure

	Beneficial use	Range of Use
1	Crop nutrient	Widespread
2	Renewable energy	Widespread; limited by installation cost
3	Soil amendment	Widespread
4	Animal feed additive	Moderate
5	Mulch	Limited
6	Livestock bedding	Limited
7	Algae production	Limited, but growing
8	Carbon sequestration	Moderate
9	Nursery pots	Limited
10	Aquaculture	Limited
11	Soil reclamation	Limited
12	Sealing ponds and dams	Limited
13	Building materials	Limited; currently largely experimental
14	Innovative figurines	Very limited
15	Paper production	Limited
16	Mushroom cultivation	Limited
17	Pesticide	Moderate
18	Plant disease control	Moderate

A typical animal unit on a confinement operation generates enough manure to produce from 20 to 55 thousand British Thermal Units (BTUs) of energy. If all the manure were collected and converted to energy, a significant amount of electrical energy could be produced. However, even when a conversion rate of 33% is assumed, each animal unit would contribute from 6 to 18 thousand BTUs each day (Table 2). Based on recent livestock and poultry inventories, almost 90 billion kWh of electricity would be produced if all manure generated on livestock farms were converted to electrical energy at a 33% efficiency of conversion.

The beneficial uses of manure are summarized concisely in Table 3. The list presented here is by no means exhaustive as the uses of manure, like those of other natural resources, changes in response to human needs and innovations. Also indicated in the table is the range of use for each specified manure benefit.

Manure Management Issues

In contrast to the wide spectrum of benefits mentioned above, manure is often associated with water and air quality problems. Manure nutrients – primarily nitrogen and phosphorus are frequently linked to water quality concerns. In addition, pathogens in manure often raise concerns about human health impacts. A USEPA survey in 2000 indicates that along with sediment, pathogens and nutrients are some of the main causes of impairment in over 35% of impaired river miles in the United States (USEPA, 2007). Pathogens were implicated as a pollutant source in more assessed river miles than any other pollutant. Bacteria were a cause of impairment in 35% of the assessed river miles that were designated as impaired. Manure from livestock agriculture was listed as one of the major sources of microorganisms that enter surface waters.

Sediment losses from agricultural lands causes various water quality problems directly and also serves as a carrier of manure nutrients and other substances that are potentially harmful to human health (Sharpley et al., 1994; Zhao et al., 2001). Nitrogen is a leading source of pollution in areas especially characterized by intensive cropland agriculture, primarily due to the use of commercial N fertilizer. Nitrogen lost through leaching and surface runoff can lead to increased nitrate levels in drinking water sources, necessitating increased water purification costs.

Phosphorus has more recently been linked to livestock operations, particularly due to imbalance between manure nutrients and nutrients required for crop uptake. In many areas of intensive livestock agriculture, P is undoubtedly the nutrient of primary concern. Phosphorus is generally adsorbed on to soil particles and released slowly over time. After sustained over-application of P on land, the assimilative capacity of soils can be exceeded leading to increased losses of soil P to downstream water bodies. Phosphorus has been directly linked to eutrophication of lakes (Sharpley and Rekolainen, 1997).

Thus, nutrients and pathogens from manure are all major contributors to the key water quality problems facing many watersheds today. One of the greatest inland water quality problems associated in some way with nonpoint source pollutants from livestock and other agricultural operations is the eutrophication of lakes and other water bodies. Eutrophication results from nutrient inputs into inland surface waters, which stimulates the growth of aquatic plants (Syers et al., 1973; Sharpley et al., 1994). A slow process of eutrophication of lakes is normal. However, accelerated eutrophication can be caused by human activity. This process, commonly termed cultural eutrophication, leads to undesirable biological activity in lakes such as algal blooms and excessive growth of other aquatic plants. Phosphorus is generally believed to be the limiting nutrient for cultural eutrophication in freshwater systems (Gibson, 1997).

Another water quality problem often linked to manure applications is hypoxia. Hypoxia is associated with reduced oxygen levels in water, primarily coastal waters, and is also caused in part by high nutrient loadings to coastal waters (CENR, 2000). Extensive hypoxia has been documented in the Gulf of Mexico and is attributed primarily to agricultural operations upstream of the gulf (CENR, 2000), especially from fertilizer use in the Midwest, but also from livestock operations. Drinking water supplies are also affected by increased erosion and nutrient losses from agriculture. Not surprisingly, in watersheds with significant livestock populations, these problems have raised public awareness of the potential water quality impacts of concentrated livestock operations.

Recent problems associated with manure management are generally perceived to be related to the changing landscape of livestock agriculture. At the national level, the number of livestock operations has declined over the past several decades, while average farm sizes have increased, in some cases, quite dramatically (USDA NASS 1982; 1992; 2002). Furthermore, increased production of livestock products such as milk and meat due to genetic improvements has potentially led to increased manure nutrient production per animal unit beyond levels that were typical about 50 years ago. These factors imply that there is more manure production on smaller land areas now than before.

MANURE PRODUCTION, STORAGE AND HANDLING

Manure is a major nutrient source for crop and forage production. In 1997, about 1.1 billion tons of manure was excreted by livestock on farms in the U.S., though a major portion of this amount was not on confined operations, and hence not collectible (PPI, 2002). A key factor that determines the portion of manure available for crop uptake is the proportion that is lost during storage and handling.

Manure production characteristics are a function of feed intake, liveweight gain and milk or egg production. In the absence of more specific information, most state and local conservation officials and livestock operators use a standard table of manure production coefficients, such as the American Society of Agricultural Engineers (ASAE) Standards (ASAE, 1995), to determine amounts of manure nutrients produced on any given farm. However, a more accurate and relatively practical measure of manure nutrient production could be obtained by using a nutrient balance model similar to Van Horn (1991, 1992), where manure N and P are estimated as the balance remaining after subtracting nutrient amounts in weight gain and livestock product sales from feed nutrient intake. More recent American Society of Agricultural and Biological Engineers (ASABE) Standards for manure production (ASABE, 2005) incorporate nutrient balance models similar to Van Horn (1992). In the absence of frequent manure testing, the disparity between actual nutrient contents and values from standards tables presents a degree of uncertainty that reduces producer resolve to account fully for manure nutrients, which leads to manure nutrient over-application (Sheriff, 2005).

The primary logistics problem related to manure nutrient management is land availability. As a whole, the above comparison between manure nutrient production and crop nutrient requirements indicates that there is enough crop nutrient demand in the U.S. to account for manure nutrients produced. However, the distribution of livestock operations presents a major challenge for manure nutrient utilization. The increased size of livestock operations in recent years without proportionately larger land areas suggests that a number of livestock operations have less land for manure utilization than was typical in the past. Furthermore, due to the bulky nature of manure, it is often difficult or costly to transport most manures to places of higher utilization potential. Thus inadequate land for manure disposal has become a limiting constraint in many livestock intensive areas regarding environmentally sustainable manure management.

In addition to the problem of land availability, crop and forage needs for manure nutrients are not adequately in sync with manure nutrient production. Due to much greater N than P losses during manure collection, handling and storage, manure N to P ratios are often no greater than 2. In contrast, the N:P ratio of crop or forage nutrient needs are generally well in excess of 5 and often as high as 8 or more. Consequently, in many livestock intensive areas P tends to be the nutrient in relative excess leading to soil P buildup where manure applications are either unchecked or based on crop N needs. Besides, in some livestock intensive areas, nutrient levels available from manure applications and commercial fertilizer use are often higher than crop nutrient requirements.

Further insight into crop nutrient needs versus manure nutrient production can be obtained by analyzing manure production characteristics and crop nutrient requirements at the county level. Kellogg et al. (2000) performed such a study based on the 1997 Agricultural Census. Their results showed that 73 out of the roughly 3,000 counties in the nation would have excess N based on manure sources versus crop needs. The number of counties with excess manure P was estimated to be greater, about 160 counties. When county level commercial fertilizer use is included, the number of counties with N and P excesses would likely increase significantly. Comparing county-level nutrient imbalances with similar measures at the national level reveals that nutrient imbalances are a regional phenomenon, being prevalent in some areas but not in others. Thus most of the imbalances would disappear if operations producing large amounts of manure nutrient were located in areas of great crop or forage nutrient demand since at the national level there is adequate crop nutrient demand for all the livestock manure produced (PPI, 2002). In particular, PPI (2002) indicates that at the national level, manure and commercial fertilizer P use exceeds uptake by harvested crops by only a small margin. A recent study by Boehlje et al. (2006) also indicates that total nutrient requirements in most states exceed livestock manure nutrient production in those states.

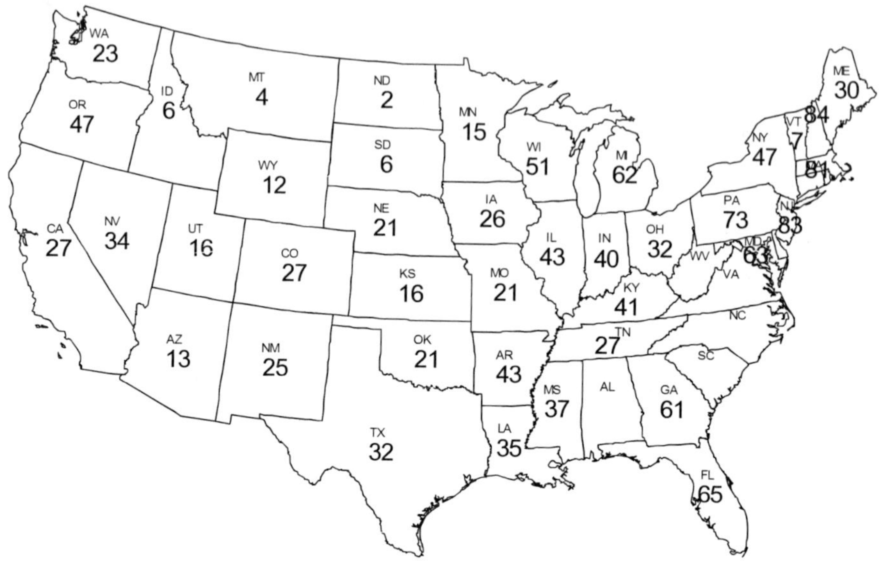

Figure 1. Percent of soils testing 40 ppm P or above in 2001[1].

In addition to nutrient losses, many environmental issues are associated with the way manure is handled before or after storage. Inappropriate manure collection and handling may result in direct discharge of pathogens and nutrients from confinement areas. Manure handling methods depend on the form in which it is collected. Solid manure is generally scraped and applied, stockpiled for later application or stored in a manure storage structure (Fulhage and Pfost, 2001). Liquid manure is collected in lagoons, settling basins, underground pits, or tanks (Fulhage and Pfost, 2001). In some cases manure is not stored at all, but simply collected and applied on application fields (Keith et al., 2000; PATS, 2001). Lack of adequate manure storage presents major challenges for manure management on a number of farms that handle manure in solid form (USDA, 2003). In the absence of manure storage structures, manure needs to be applied on land soon after collection or stockpiled on fields for spreading at a later date. In either case, a significant potential exists for nutrient losses during rainfall events. This practice also increases the chances of pathogen transfer from fresh manure to surface water resources.

[1] This figure was based on data from PPI (2001). Data not available for AL, NC, SC, VA, and WV.

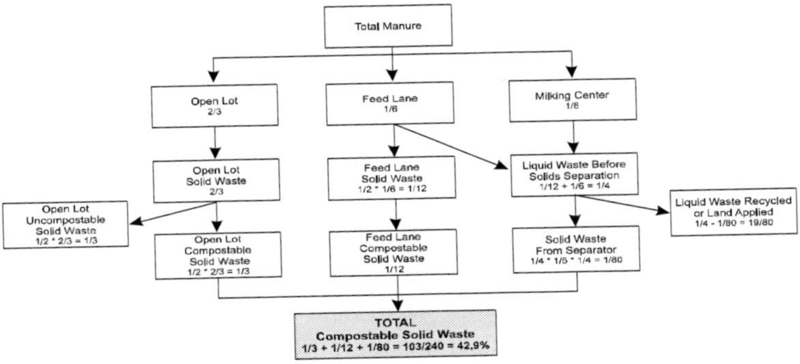

Figure 2. Fractional dairy manure composition by source and use for open-lot dairies in the UNBRW[2]

Manure Nutrient Uncertainty

One of the major problems compounding the issue of manure management is the uncertainty associated with its nutrient composition. There is a degree of uncertainty associated with nutrient loss, especially for N, at each phase of the manure generation, storage, handling, and disposal cycle (NRC, 1993). Due to this uncertainty, most producers understandably have relatively little confidence in manure nutrient content. This uncertainty is usually blamed in part for the limited acceptance of manure nutrient crediting by livestock operations. In order to achieve greater rates of adoption of improved manure nutrient management practices by livestock producers, these sources of uncertainty need to be addressed.

In addition to uncertainties about manure nutrient content, there is also a degree of uncertainty about the plant availability of manure nutrients. It is generally held that the organic components of manure N and P are less readily available for plant uptake than the inorganic component of manure nutrients or commercial fertilizer sources. Consequently, in many cases, producers are allowed to apply manure at higher rates in order to ensure adequate plant-available nutrient levels in the soil. However, when overly conservative assumptions are made about plant availability of manure nutrients and subsequent mineralization and increased plant availability is ignored, producers are more likely to overapply manure and environmentally detrimental losses of nutrients could occur.

[2] Data in this figure were based in part on Sweeten (1998).

Soil Nutrient Buildup

Besides the problem of inadequate land in some livestock intensive areas, consistent overapplication of manure nutrients over a long period of time has led to an overabundance of plant nutrients in the soils in livestock intensive areas, so that some of the available land is unsuitable for manure disposal. This overabundance is particularly true of P, and in many cases, decades of crop or forage uptake without additional P application is required for soils to become suitable for manure disposal (Sharpley et al., 1994; Sharpley and Rekolainen, 1997). Figure 2 shows the percentage of soils by state that tested at 40 ppm Bray P-1 equivalent or higher in 2001. Soil P levels above 40 ppm are considered high or very high in most situations, suggesting that little or no supplemental P fertilizer is needed on these soils. Thus Figure 2 suggests that in some areas most soils are already so high in P that the potential for P loss from further manure applications is great.

Soil P buildup has been an issue of considerable frustration for many producers, policymakers, and stakeholders. This buildup implies that nonpoint source P loads to surface water resources may not improve appreciably even after years of implementing the most effective best management practices (BMPs). This observation, which is sometimes erroneously perceived as an indication of BMP ineffectiveness, is largely due to the fact that much time is needed for P to be mined out of the soil profile.

Problems on Grazed Pastures

In spite of the movement toward larger and more confined animal feeding operations (AFOs), many livestock farms are still pasture-based operations. Whereas the focus of environmental issues relating to livestock has been on confined operations, recent studies have shown that pasture operations are also an appreciable source of potential nonpoint nutrient pollution of surface waters especially where livestock have direct access to streams (Correl, 1996; Belsky et al., 1999; Hubbard et al., 2004).

Three key factors account for nutrient losses from grazed pastures: denudation, excessively high stocking densities, and excessive use of inorganic fertilizers (McNitt et al., 1999). Denudation refers to the depletion of vegetative cover in portions of pasture as livestock congregate, usually around feeding and watering areas. Denudation results from improper management of livestock movement on pasture. Loss of vegetative cover and concomitant increased manure deposition rates on denuded patches lead to concentrated runoff from these areas during rain events. Excessively high stocking densities also lead to impairment of downstream waters. When stocking densities are excessively high and supplemental feed is still used, manure P deposition rates exceed forage P uptake. Furthermore, some pastures also receive more inorganic fertilizers than needed because manure nutrients are not adequately accounted for. This is partly justified by the fact that manure is deposited unevenly, particularly on unmanaged pastures, thus requiring some fertilizer use in spite of the levels of manure nutrient already going on the pasture.

In addition to the three factors outlined above, livestock grazing in riparian areas can cause destabilization of creek banks, and even deposit manure directly into the beds of rivers and streams (Belsky et al., 1999). The extent of such impairment has, however, not been adequately quantified in research studies.

Other Impacts of Manure

While water quality issues have received far greater attention in watersheds where manure has been linked to environmental concerns, odor and other nuisance pollutants associated with manure management areas have also received attention (Huang and Miller, 2006; Schiffman et al., 1995; Khuder et al., 2007; Correll, 1996; Webb et al., 2005; Willinger, 1974). Consequently, areas of intensive manure production and disposal have been shown to impact neighboring property values (Herriges et al., 2003; Abdalla, 2003; Palmquist et al., 1997; Nehring et al., 2006; Vukina and Wossink, 2000). Furthermore, USEPA data show that methane emissions from livestock manure management areas are a small but noticeable portion of total U.S. methane emissions (USEPA, 2005; 2007).

Issues with BMP Implementation

A large number of practices have been proposed as BMPs to help alleviate the negative environmental externalities often associated with livestock agriculture. One of the issues limiting the effects of these practices is the potential for cross-media transfers (Osei and Babcock, 1996; Aillery et al.,

2005). A cross-media transfer occurs if a nutrient or pollutant load to one environmental medium (e.g., water) is reduced by use of a practice, while the same practice leads to an increase in the load of the same nutrient in another environmental medium (e.g., air). A typical example of this is the case of lagoons or waste storage ponds. Use of these structures leads to significant reduction in N losses to surface waters that would have occurred through direct discharge. This reduction occurs because, without storage structures, manure nutrients would generally be washed off directly to downstream waters during storm events. However, N losses to the atmosphere through volatilization are increased when manure is stored in lagoons. In contrast to the use of lagoons, manure storage tanks or pits would preserve a greater proportion of the N because they are often covered.

Another complication with BMP implementation relates to sunk costs. Over the years, as more information is obtained about effectiveness of conservative practices, farmers are encouraged to adopt new BMPs. However, in many cases, migration to new practices means a loss of the economic use of a number of equipment, facilities, or structures. For example, suppose a two-stage lagoon was installed on a farm to hold liquid manure and wastewater runoff. If a few years after installation, conservation improvements are made on the farm requiring that all manure be handled in solid form, then most likely the economic use of the lagoon system would be diminished. Due to such irreversible investments, the efficacy of BMPs needs to be ascertained before producers are encouraged to adopt them. Another reason for the occurrence of sunk costs in environmental practices for agriculture is that as farms change in size, many structures and facilities need to be redesigned.

Options for Effective Manure Management

A variety of options are available for livestock operators to deal with the problems and issues encountered in relation to manure management. During initial federal, state, and local government attempts at encouraging environmental stewardship on livestock operations, producers were encouraged to adopt individual practices. More recently, due to a greater awareness of the synergies and interrelations between various practices, efforts have been directed towards comprehensive suites of management and structural practices. The following sections present the primary options available to livestock operations in several categories, most of which are identical to some of USDA's Comprehensive Nutrient Management Plan (CNMP) elements. Some of the options listed below for manure management parallel the beneficial uses of manure listed above. However, additional discussion is given below of the pros and cons of each option, and the economic and environmental implications.

Table 4. Issues relating to manure management

	Issue	Causes
1	Direct discharge from confinement areas	Lack of containment structure; lots not scraped clean
2	Methane and other air pollution emissions	Manure left standing on lots
3	Runoff from manure application fields	Manure not incorporated/injected into soil
4	Toxic spills from lagoons and ponds	Lagoons not properly sealed; not dewatered as needed
5	Soil nutrient buildup	Overapplication of manure
6	Eutrophication of receiving inland waters	Runoff of manure nutrients, particularly phosphorus due to soil buildup or manure erosion
7	Imbalance of manure nutrients for crop needs	Natural result of manure losing substantial nitrogen during storage and handling
8	Adverse impacts on neighboring property values	Odor and other nuisances that limit buyer interest in properties close to manure management areas
9	Streambank/creek destabilization	Unrestricted/unmanaged livestock access to riparian areas
10	Direct deposition of manure in creeks	Unrestricted/unmanaged livestock access to riparian areas
11	Denudation of pastures	Overgrazing; unmanaged grazing
12	Manure nutrient uncertainty	Unknown impacts of weather and multiple biological processes on fate and transport of manure nutrients

Feed Management

Feed management options have a direct influence on manure production characteristics. Manure nutrient content can be altered by appropriate livestock ration modifications and feed waste control. Careful feed management will help minimize the burden of overall manure management on livestock farms.

Feed ration modifications are playing an increasing role in the management of manure nutrients on AFOs. Over the past few decades, a number of research studies have shown that significant changes in manure nutrient production can be obtained by altering feed rations (Van Horn et al., 1996; Satter and Wu, 1999; Bosch et al., 1998). Of particular interest is ration modifications to reduce the amount of manure P generated by livestock and poultry. To meet this objective, studies have suggested dietary P reduction for livestock, particularly dairy cows, and use of feed additives, primarily enzymes, to increase P digestive efficiency thereby decreasing dietary P intake with consequent reduction in P excretion for poultry and swine. A variety of these options were explored in the NPP's watershed studies.

Reduced P diets are particularly relevant for dairy herds. Studies have shown unequivocally, that confined lactating dairy herds are often fed in excess of the levels of dietary P required for maintenance and lactation (Van Horn et al., 1996; Morse et al., 1992). Justification for higher than necessary dietary P has traditionally been based on the need to maintain reproductive ability and milk production and quality (Chase, 1998). However, recent studies indicate that dietary P can be reduced to recommended levels with no adverse effect on the dairy herds or milk production (Wu and Satter, 2000; Chase, 1998). This is a very cost effective option because reduced dietary P levels translate directly into lower manure P excretion, generally in proportion to the reduction in ration P. Reduction in dietary P also yields economic benefits to livestock farmers. Several studies indicate that lower dietary P levels translate proportionately to lower feed costs, primarily due to reduced expenses on supplemental P sources (Rotz et al., 2002; Keplinger and Abraham, 2002).

Another feed management option is the use of the phytase enzyme as a feed additive for swine and broiler operations. Phytase increases the proportion of phytate P assimilated by hogs and poultry, thus reducing manure P excretion from these animals (Honeyman, 1993; Radcliffe et al., 1998). Because of the higher digestive efficiency, phytase supplementation means that lower dietary P intake is needed. In general, hog producers who use the phytase enzyme do not experience an appreciable change in livestock feed costs (Bosch et al., 1998; Progressive Farmer, 1998). On the other hand, for typical levels of phytase use, manure P excretion decreases by about 30 to 40% on average (Honeyman, 1993). Similar environmental impacts have been documented for broiler operations. Reduced manure P excretion may also lead to lower manure application costs if manure needs to be applied at the P agronomic rate (Bosch et al., 1998).

Manure and Wastewater Handling and Storage

Initial efforts to control nutrient losses from livestock operations focused on preventing direct discharge of manure and manure nutrients from confinement areas to receiving water bodies. To this end a number of structural and managerial practices were suggested to enhance manure collection, storage, and transport on farms. EPA's "no discharge" requirement for concentrated animal feeding operations means that all manure and wastewater must be contained on the farm and disposed of appropriately.

A number of structures and practices enable livestock operators to meet the "no discharge" requirements. Manure storage structures prevent direct discharge of manure nutrients and pathogens and allow for flexible timing of manure use. Runoff containment structures collect runoff from open lot and other confinement areas. Diversions keep clean water from coming into contact with wastewater. Sediment basins and filter strips allow solids to settle out of runoff and provide for treatment of suspended solids, nutrients and other contaminants. In addition, appropriate mortality disposal and on-farm manure treatments and transport mechanisms all serve to minimize direct discharge of manure and wastewater to surface water resources.

Substantial progress has been made over the years through the use of these practices. Where these options have been put in place, direct manure and nutrient loss from confinement areas are rarely a problem. However, in some regions of the country, a number of essential structural and management practices are noticeably absent on AFOs. Of particular significance is the absence of adequate manure storage structures on many small operations. As a result, manure is of necessity applied on cropland very frequently, making it more likely for some manure applications to occur just prior to a rainstorm or on frozen ground. A study on CNMP implementation costs indicates that about 29% of small dairies in the North Central and Northeastern states of the U.S. (over 15,000 farms) have no manure storage (USDA, 2003). Manure storage structures were identified as key components of CNMPs on those farms.

Land Application Options

Land application methods are some of the most important options for reducing agricultural nonpoint source pollution (Pratt et al., 1997). While initial efforts at reducing N and P losses from livestock operations focused on structural controls, manure utilization on cropland has received far greater attention in recent years (Pratt et al., 1997). Four aspects of manure nutrient use are essential for controlling nonpoint nutrient losses from AFOs: manure application rates, nutrient accounting, mode of nutrient application, and timing of nutrient applications.

Many intensive livestock operations are in areas of minimal crop nutrient demand. In such areas, most producers see manure management as more of a disposal problem than a nutrient management opportunity. Consequently, when permits or management plans for livestock operations did not give due attention to land application methods, operators tended to apply manure at rates exceeding nutrient needs on crop or forage fields and pasture. This practice led to soil P buildup after many years of sustained over-application (Sharpley and Rekolainen, 1997; Pratt et al., 1997).

To prevent these problems, more recent livestock operation permits and management plans specify manure applications at crop agronomic uptake rates (Pratt et al., 1997). Initially, most regulatory agencies and producers were content to have manure applications at crop N agronomic rates even though, as discussed earlier, crop N/P ratios are generally higher than manure N/P ratios. The N rate was used partly because of the prevailing notion at the time, that most soils would retain increasing P levels without releasing them to receiving water bodies (Sharpley et al., 1994) and also because in many livestock intensive areas, manure applications at P agronomic rates are more expensive. However, current understanding indicates that soils over-saturated with P release significant amounts of bioavailable P to receiving waters, causing eutrophication problems in many areas (Sharpley et al., 1994). Consequently, current manure management plans place increasing emphasis on soil testing for P and applications at crop P uptake rates, when soil test P levels reach certain thresholds. As mentioned earlier, P uptake rates are generally much lower than N uptake rates.

Lower manure application rates imply that manure needs to be applied over greater acreage than before. If manure needs to be applied at the P uptake rate and was previously applied at the N rate, the increase in acreage required is in proportion to the difference between the crop N/P ratio and the manure N/P ratio. The water quality benefits of manure applications at P uptake rather than N uptake or higher rates are supported both by intuition and by results from several watershed studies (Osei et al., 2000b; Keith et al., 2000; Gassman et al., 2006) and large-scale regional simulations (Osei et al., 2008). However, in some instances, a different issue is presented. What happens if the soils that now receive manure but did not in previous years, are more susceptible to erosion? That case would indicate that the benefits of improved manure application rates might be offset to some degree by higher erosion rates, particularly if the new fields were not previously cultivated and did not previously receive any significant nutrient applications.

The economic impacts of lower manure application rates vary widely from one production region to another. Financial impacts on producers depend largely on the availability of crop nutrient (particularly P) demand in the vicinity of the livestock operation. In crop intensive areas of the Midwest where there is great crop demand for manure nutrients, more producers benefit financially from lower manure application rates and judicious fertilizer use (Keith et al., 2000). On the contrary, many AFO owners incur moderate to significant costs in areas with low crop nutrient demand (Osei et al., 2000b; Pratt et al., 1997).

Applying manure at crop agronomic (preferably P) uptake rates is one key step in appropriate nutrient management on AFOs, but this in itself does not sufficiently constitute appropriate nutrient accounting. Appropriate nutrient accounting certainly means that nutrient levels in manure should be considered when determining supplemental inorganic fertilizer use. In addition nutrient accounting also requires that soil nutrient levels and reasonable legume N sources are considered. Generally, P-based manure applications would require supplemental N fertilizer for adequate crop nutrition (Osei et al., 2000b), which follows directly from the values in Tables 1 and 2. Appropriate manure application rates together with adequate nutrient accounting help ensure that soil nutrient levels remain within acceptable limits.

Once manure and fertilizer applications have been determined at rates that avoid nutrient buildup in soils, nutrient losses from agricultural fields depend to some extent on the mode of application. Surface application of liquid or solid manure results in 20 to 30% N volatilization losses. In addition, nutrients left on the soil surface are more susceptible to erosion during a rain event. Solid manure can be incorporated to minimize surface runoff nutrient losses. For a similar purpose, liquid manure can be injected. Usually, soluble nutrient losses are reduced when manure is incorporated or injected, sediment and sediment-bound nutrient losses may increase depending on how the manure is incorporated (e.g., Romkens et al., 1973; Mueller et al., 1984; Sharpley and

Rekolainen, 1997; Zhao et al., 2001; Little et al., 2005). The impact on total nutrient losses thus depends on the proportions of soluble and sediment-bound nutrient fractions of manure nutrients (Osei et al., 2003a). Incorporation and injection procedures also result in much lower N volatilization losses. Typically, volatilization losses are no more than 5% when manure is placed below the soil surface.

Due to decreased N volatilization losses with manure incorporation or injection, application rates can and should be adjusted (reduced) to account for lower N losses, particularly if N is the target nutrient. Notwithstanding, in most cases, with nutrient accounting farmers will see a reduction in commercial fertilizer expenses with injection or incorporation, whether manure is being applied at the N or the P agronomic rate.

Manure application timing is also very crucial to nutrient loss prevention. Research has shown that most of the P loss from crop fields in any given year occurs in one or two storm events (Sharpley et al., 1994). Thus manure application just prior to a storm event will lead to increased nutrient losses than if the application had been made at other times. On farms with adequate manure storage structures, there is greater flexibility in scheduling manure applications. Manure storage allows farmers to refrain from manure applications during winter months. Especially in colder climates, manure is applied very frequently during winter on farms with inadequate manure storage, which increases the potential for nutrient loss when snow melt or spring rains occur.

Land Management Options

Land management practices include a variety of managerial and structural options that influence soil erosion and losses of manure nutrients and other agricultural chemicals and sediment from farmland. These practices are more important for cultivated cropland areas than for pastureland.

Conservation tillage options (no-till, reduced tillage) have been shown to reduce sediment losses from cropland by as much as 50% relative to conventional tillage methods (Gassman et al., 2006). Economic impacts are usually mixed. In general, there is either a small gain or a small loss in farm net returns when conservation tillage options are practiced relative to conventional tillage (Gassman et al., 2006). Conservation tillage options entail fewer field operations and use of special machinery to minimize inversion of topsoil. Enhanced vegetative cover is the primary reason for reduced sediment losses. Consequently losses of sediment-bound nutrients are also lower, which is particularly important in intensive cropland areas. However, there is also increased infiltration, which may lead to higher soluble nutrient losses, especially N, through subsurface flow. Thus the net effect is dependent on the relative proportions of soluble and sediment-bound nutrients and the nutrient(s) of water quality concern.

Contour farming on high slope cropland can also reduce soil erosion, much like conservation tillage. Contouring reduces the slope length and hence the velocity of runoff water thus increasing infiltration rates. The environmental effects are similar to those of conservation tillage practices. For the producer, contouring generally entails a small reduction in field efficiency for operations performed along contours, which may lead to a small cost increase per acre of contoured land (Gassman et al., 2006; Keith et al., 2000). Terracing functions in a manner similar to contouring, and has even greater benefits on water quality. However, in the absence of cost-share funding, the financial burdens on farmers are much greater with terracing.

Filter and buffer strips control the movement of sediment and agricultural nutrients from cropland. Filter strips are used at the downslope edge of manure application fields in widths that vary depending on the nutrient to be controlled, slope, and crop cover. They are more effective for the control of sediment and sediment-bound nutrients than soluble nutrients in surface runoff (Dosskey, 2001). While maintenance of filter strips generally does not entail significant costs, a major economic impact relates to the fact that some land is taken out of production (Keith et al., 2000; Gassman et al., 2006). The impact of vegetative filter strips on reducing sediment and nutrient losses from cropland varies greatly, and this is still the subject of much study (Dosskey, 2001).

Buffer strips can also be effective in reducing sediment losses from cropland. They are used in a variety of ways in livestock operations. Contour buffer strips are very effective for the control of sediment and organic and soluble nutrients alike (Keith et al., 2000). Buffer strips are placed in between wider strips of cultivated crops. Buffer acreage may contain forages that can be harvested for on-farm use or sale or they may simply contain perennial grasses that have little or no productive use. The economic impact of buffer strips is highly dependent on the value of productive use forgone (Keith et al., 2000). For instance, buffer strip acreage in a highly intensive corn and soybean production area entails greater net cost to producers than strips installed in low productive land. In many cases producers receive government payments to offset some of these costs. Other land management practices that help reduce erosion and nutrient losses from cropland include, grassed waterways (Gassman et al., 2006), field borders, wetlands, riparian zone management, buffer zones, and use exclusion.

Pasture Management Options

Many livestock operations, particularly beef and dairy farms, still operate pasture-style systems. For these operations, careful grazing management can result in significant gains for producers and the environment in areas where uncontrolled grazing activities have led to higher than necessary nutrient losses from pasture. In intensive grazing management systems, livestock are rotated from one grazed sub-pasture or paddock to another. Careful management of livestock movement and grazing minimizes the incidence of denuded areas on pastures. These denuded areas can be a significant cause of nonpoint nutrient pollution as they were in the Lake Fork Reservoir watershed in Texas (McNitt et al., 1999).

Intensive grazing management also enhances forage use, thereby providing some financial benefits to producers. However, the overall impact on farm profits is not always obvious. For instance, increased forage intake for dairy cows on intensively grazed pasture usually means lower milk yields unless some cost increasing adjustments are made in livestock feed rations. Generally, feed costs are lower on intensive grazing systems, but revenues from livestock products are also somewhat lower, as are maintenance and repair expenses. The overall impact depends on the managerial effectiveness of livestock operators, but on small to medium sized operations, there is usually a small profit increase compared to uncontrolled grazed pasture operations or confined livestock housing systems (McNitt et al., 1999).

Aside from intensive grazing systems, cattle movement on pasture may also be managed around feeding areas to prevent denudation by moving the cows from one loafing lot to another and maintaining grass cover on the lots. This practice allows producers to maintain existing housing systems and still prevent nutrient losses resulting from denuded patches (McNitt et al., 1999). Finally, livestock access to creeks and stream banks can be controlled to prevent direct deposition of manure in surface waters or stream bank destabilization.

Other Utilization Options

There are a number of alternative uses for livestock manure, in addition to its primary use as a nutrient source for crop and forage production. Alternative options include composting manure for agricultural or nonagricultural uses, power generation, feedstock for livestock rations, and various agricultural or manufacturing processes that develop high valued products from manure nutrients.

The amount of manure available for alternative uses depends on how manure is handled on the farm. Figure 3 shows the fractional composition of dairy manure by source and use for an open-lot dairy in the UNBRW. The figure indicates that, whereas most of the manure was handled in solid form, only about 43 percent of manure generated on a typical dairy farm in that watershed was compostable. Thus efforts toward enhancing other utilization options must be accompanied by adjustments in on-farm manure handling systems.

Composting renders solid manure more marketable and increases the range of its uses. However, composting often entails significant costs, as is the case in the North Bosque River watershed (Osei et al., 2000b; Pratt et al., 1997), and may not be a viable alternative if ready markets are not available for the finished product. Composting dairy manure costs about $10 per ton of incoming manure, though in some areas the finished compost can be sold profitably. Where viable, composting for end uses outside impacted watersheds is a sure way of moving manure nutrients outside areas of heavy soil build up.

Alternative technologies exist for transforming manure into useable energy. In recent years, anaerobic digestion system shave been installed on a number of farms with varied success rates. Economic analyses conducted on these projects indicate that anaerobic digestion and other power generation technologies are feasible primarily for larger sized operations (Lazarus and Rudstrom, 2007; Osei and Tanter, 2007; Dagnall, 1995). Capital requirements for these technologies are often too prohibitive for small operations without external funding support.

Summary of Options for Controlling Pollution from Livestock

Table 5 summarizes the major problems and issues facing livestock operations today in relation to water quality, along with potential options that address those issues. The key problems highlighted are direct discharge from confinement areas, soil nutrient buildup and erosion from manure application fields, and soil nutrient buildup, denudation, and creek and stream impairment in relation to grazing livestock. The efficacies of these options depend on many factors specific to the region of concern and the management style of the farm. Nonetheless, the use of practice combinations as suggested in CNMPs holds promise for reduction in sediment and nutrient losses.

Table 5. Manure management options

	Option	Pros/Cons
1	Feed ration modifications	One of least expensive ways to control manure nutrient levels at the source
	a. Phytase	Cost-effective enzyme additive that increases bioavailability of phosphorus and reduces manure phosphorus concentrations. Applicable to monogastric animals
	b. Reduced phosphorus diets	Cost-effective way to reduce manure phosphorus excretion and management costs for ruminant animals
2	Manure storage structures	Very expensive; reduces/eliminates direct discharge of manure, but also increase nitrogen losses prior to land application
3	Manure handling	Minimizes odor and direct discharge of manure; may increase nitrogen losses and imbalance of manure for crop needs
4	Judicious manure application rates	One of most important management options; reduces nutrient losses in runoff
	a. Full accounting/credit for manure nutrients	Reduces soil nutrient buildup; reduces farm costs; may reduce crop yields if expected manure nutrient levels incorrect
5	Manure incorporation/injection	Reduces nitrogen volatilization losses and runoff losses in soil erosion
6	Timing of manure and other nutrient applications	Reduces potential for high nutrient losses in soil erosion; could hinder timeliness of farm operations due to unforeseen weather events
7	Conservation tillage	Reduces erosion and consequent manure nutrient losses; cost-effective
8	Land management	Reduces erosion and sediment-bound manure nutrient losses; can be costly
9	Composting	Expensive; makes manure more marketable; reduces threat of pathogens; reduces weed seeds and provides

		more consistent quality of manure
10	Energy generation	Expensive, largely infeasible for small operations; generates useful bioenergy from manure; captures methane, a greenhouse gas
11	Manure banks/markets	Provides more visibility for manure; enhances manure transfers to nutrient deficit areas.
12	Processed manure products	Makes manure more marketable
13	Controlled livestock access to creeks	Reduces streambank/creek destabilization and direct deposition of manure in riparian areas.
14	Managed grazing	Reduces nutrient runoff from pastures; very cost-effective
15	Reduced pasture stocking densities	Reduces nutrient runoff from pastures; can be quite costly if land area is inadequate

Conclusions

Manure, particularly animal manure, is a very important resource that is often overlooked and even regarded as a waste in some areas. However, manure has many beneficial uses in addition to its primary use as a holistic crop nutrient source. Among the many benefits of manure, it can be used as feedstock for energy generation, works well as a soil conditioner, is an animal feed additive, can be used as mulch in farms or gardens, is used for livestock bedding when processed, and in algae production, and can even be used in the manufacture of paper products.

Notwithstanding the many benefits of manure, it can be a source of human health and environmental problems when used inappropriately. In this chapter, we outlined the benefits of manure, the key issues associated with its use, and the variety of options available for addressing those issues and putting manure to beneficial use. While manure includes livestock excreta as well as crop biomass the focus in this chapter was on animal manure.

On-farm manure management issues are best addressed in a holistic manner. On confined livestock operations, attention needs to be paid to manure application fields, and not only the confinement areas. Efforts to reduce nutrient losses from manure application fields will entail some financial impacts, which depend largely on the availability of crop nutrient demand in the vicinity of the livestock operation.

Pasture operations can also improve their management styles to reduce nutrient losses from grazed pastures. Improved grazing management minimizes denudation, appropriate stocking densities reduce soil nutrient buildup, and reduced commercial fertilizer use in accordance with manure nutrient deposition on pasture also serves to enhance the environmental stewardship of grazed pastures.

Significant progress has already been made by producers, stakeholders, and government agencies in addressing manure management issues. Furthermore, current emphasis on adaptable suites of management practices should help achieve water quality goals. Future research should give priority to issues such as minimizing manure nutrient uncertainty, exploring alternative uses of manure, improving manure marketing options, and understanding and quantifying the role of grazed pastures in water quality.

REFERENCES

Adhikari, M; Paudel, KP; Martin, Jr NR; Gauthier, WM. Economics of dairy waste use as fertilizer in central Texas. *Waste Management*, 2005, 25, 1067-74.

Aillery, M; Gollehon, N; Johansson, R; Kaplan, J; Key, N; Ribaudo, M. Managing manure to improve air and water quality. Economic Research Service Report 9. U.S. Department of Agriculture, *Economic Research Service*. Washington DC. September 2005.

Albihn, A; Vinneras, B. Biosecurity and arable use of manure and biowaste – treatment alternatives. *Livestock Science*, 2007, 112, 232-9.

Araji, AA; Abdo, ZO; Joyce, P. Efficient use of animal manure on cropland – economic analysis. *Bioresource Technology*, 2001, 79, 179-91.

Araji, AA; Stodick, LD. The economic potential of feedlot wastes utilization in agricultural production. *Biological Wastes*, 1990, 32, 111-24.

Arnold, JG., Srinivasan, R; Muttiah, RS; Williams, JR. Large area hydrologic modeling and assessment; Part I, Model development. *Journal of the American Water Research Association*, 1998, 34(1), 73-89.

Asabe. Manure production and characteristics. *American Society of Agricultural and Biological Engineers*. Publication D384.3, March 2005.

Asae. Manure production and characteristics. ASAE Data: ASAE D384.1, ASAE Standards. *American Society of Agricultural Engineers*, 1995.

Ashraf, M; Christensen, RL. An analysis of the impact of manure disposal regulations on dairy farms. *American Journal of Agricultural Economics*, 1974, 56, 331-6.

Barron, Teddi. From cow chips to cow barns. Inside Iowa State. May 19, 2000. Available at *http: / /www. Iastate .edu /Inside /2000/0519 /cowchips. html*. Last accessed, 2000, October 28, 2009.

Bazen, EF; Fleming, RA. An economic evaluation of livestock odor regulation distances. *Journal of Environmental Quality*, 2004, 33, 1997-2004.

Belsky, AJ; Matzke, A; Uselman, S. Survey of livestock influences on stream and riparian ecosystems in the western United States. *Journal of Soil and Water Conservation. First Quarter*, 1999, 419-431.

Blackburn, JW. Effect of swine waste concentration on energy production and profitability of aerobic thermophilic processing. *Biomass and Bioenergy*, 2001, 21, 43-51.

Boehlje, M; Gray, A; Mark, T. The growth potential for the Indiana livestock industries. Purdue University, *Department of Agricultural Economics*, Staff Paper #06-06, April 2006, West Lafayette, IN.

Boland, MA; Foster, KA; Preckel, PV. Nutrition and the economics of swine management. *Journal of Agricultural and Applied Economics*, 1999, 31, 83-96.

Boland, MA; Preckel, PV; Foster, KA. Economic analysis of phosphorus-reducing technologies in pork production. *Journal of Agricultural and Resource Economics*, 1998, 23, 468-82.

Bosch, DJ; Pease, JW. Economic risk and water quality protection in agriculture. *Review of Agricultural Economics*, 2000, 22, 438-63.

Bosch, DJ; Wolfe, ML; Knowlton, KF. Reducing phosphorus runoff from dairy farms. *Journal of Environmental Quality*, 2006, 35, 918-27.

Bosch, DJ; Zhu, M; Kornegay, ET. Economic returns from reducing poultry litter phosphorus with microbial phytase. *Journal of Agricultural and Applied Economics*, 1997, 29, 255-66.

Bosch, Darrell, J., Minkang Zhu, ET. Kornegay, Net returns from microbial phytase when crop applications of swine manure are limited by phosphorus. *Journal of production agriculture*, 1998, Vol. 11. No. 2 205-213.

Bottcher, AB; Tremwel, TK; Campbell, KL. Best management practices for water quality improvement in the Lake Okeechobee watershed. *Ecological Engineering*, 1995, 5, 341-56.

Bourdeau, P; L'Hermite, P. An EEC. viewpoint on animal waste disposal. Philosophical Transactions of the Royal Society of London, *Series B: Biological Sciences*, 1982, 296, 333-9.

Brown, BB; Yiridoe, EK; Gordon, R. Impact of single versus multiple policy options on the economic feasibility of biogas energy production: *swine and dairy operations in Nova Scotia. Energy Policy*, 2007, 35, 4597-610.

CAST. Summary: Integrated Animal Waste Management. Council for Agricultural Science and Technology. Available at *http://www.cast-science.org/websiteUploads/pdfs/anwa_is.pdf.* Last accessed, 1996, October 28, 2009.

CENR. 2000. Integrated assessment of hopoxia in the northern Gulf of Mexico. National Science and Technology Council Committee on Environment and Natural Resources, Washington, DC. May 2000. Available at *http://oceanservice.noaa.gov/products/pubs_hypox.html.*

Chalker-Scott, Linda. Horticultural myths: compost teas: examininig the science behind the claims. Master Gardener. Winter 2007. Available at *http://www.MasterGardenerOnline.com.*

Chase, LE. Phosphorus nutrition of dairy cattle: In Mid-South Ruminant Nutrition Conference Proceedings, May 7-8, 1998, Ellen Jordan, ed.: *The Texas Animal Nutrition Council & the Texas Agricultural Extension Service*, Dallas, Texas, 1998.

Cheung YH; Wong, MH. Properties of animal manures and sewage sludges and their utilisation for algal growth. *Agricultural Wastes*, 1981, 3, 109-22.

Clark, S. ZEROS 'System' Technology. Technology Due Diligence Report. August 2006. ZEROS Incorporated. Available from: URL: http://www.zerosinc.com/ Due_Diligence.pdf (accessed 22 August 2007).

Collins, AR; Basden, T. A policy evaluation of transport subsidies for poultry litter in West Virginia. *Review of Agricultural Economics*, 2006, 28, 72-88.

Correll, DL. Environmental impacts of pasture systems on surface water quality. In: RE; Joost, CA. Roberts, editors. Nutrient Cycling in Forage Systems. *Potash and Phosphate Institute and the Foundation Agronomical Research*, Manhattan, KS, 1996.

Cowell, DA; Apsimon, HM. Cost-effective strategies for the abatement of ammonia emissions from European agriculture. *Atmospheric Environment*, 1998, 32, 573-80.

Dagnall, S. UK strategy for centralized anaerobic digestion. *Bioresource Technology*, 1995, 52, 275-80.

De Haan FAM, Van der Zee SEATM. Soil protection and intensive animal husbandry in the Netherlands. *Marine Pollution Bulletin*, 1994, 29, 439-43.

Delgado, C; Rosegrant, M; Steinfeld, H; Ehui, S; Courbois, C. Livestock to 2020, the next food revolution. Food, Agriculture, and the Environment Discussion Paper 28. International Food Policy Research Institute, *Food and Agriculture Organization of the United Nations*, and International Livestock Research Institute. Available from: URL: http://www.ifpri.org/2020/dp/dp28.pdf.

Dietz, FH; Hoogervorst, NJP. Towards a sustainable and efficient use of manure in agriculture: the Dutch case. *Environmental Resource Economics*, 1991, 1, 313-32.

Dobbs, TL; Pretty, JN. Agri-environmental stewardship schemes and 'multifunctionality'. *Review of Agricultural Economics*, 2004, 26, 220-37.

Donham, KJ; Wing, S; Osterberg, D; Flora, JL; Hodne, C; Thu, KM; et al. Community health and socioeconomic issues surrounding concentrated animal feeding operations. *Environmental Health Perspectives*, 2007, 115, 317-20.

Dosskey, Michael, G. Toward quantifying water pollution abatement in response to installing buffers on crop land. *Environmental Management*, 2001, 28 (No. 5), 577-598.

Feinerman, E; Bosch, DJ; Pease, JW. Manure applications and nutrient standards. *American Journal of Agricultural Economics*, 2004, 84, 14-25.

Fleming, RA; Babcock, BA; Wang, E. Resource or waste? The economics of swine manure storage and management. *Review of Agricultural Economics*, 1998, 20, 96-113.

Fleming, RA. The economic impact of setback requirements on land application of manure. *Land Economics*, 1999, 75, 579-91.

Flores, Alfredo. 2008. "Organically Grown Pecans Show Significantly Higher Yields vs Conventionally Grown!" Available at http://www.ghorganics.com/High Yielding Organic Pecan Trees.htm.

Forster, DL. Simulated beef feedlot behavior under alternative water pollution control rules. *American Journal of Agricultural Economics*, 1975, 57, 259-68.

Freschet, GT; Masse, D; Hien, E; Sall, S; Chotte, JL. Long-term changes in organic matter and microbial properties resulting from manuring practices in an arid cultivated soil in Burkina Faso. Agriculture, *Ecosystems and Environment*, 2008, 123, 175-84.

Fulhage, CD; Pfost, DL. Economic considerations for beef manure management systems. MU Guide. MU Extension, University of Missouri-Columbia. Report No. EQ 392. April 2003.

Garrett, Howard. 2009. The Dirt Doctor: Organic Recipes Homemade: Compost tea. Available at http: // www. Dirtdoctor .com/ organic/garden/view_question/id/204/. Last visited October, 28, 2009.

Gassman, PW; Osei, E; Saleh, A; Hauck, LM. Application of an environmental and economic modeling system for watershed assessments. *Journal of the American Water Resources Association*, 2002, 38, 423-38.

Gassman, PW; Osei, E; Saleh, A; Rodecap, J; Norvell, S; Williams, JR. Alternative practices for sediment and nutrient loss control on livestock farms in northeast Iowa. Agriculture, *Ecosystems and Environment*, Vol. 117 (November 2006), 135-144.

Gerber, P; Menzi, H. Nitrogen losses from intensive livestock farming systems in Southeast Asia: a review of current trends and mitigation options. *International Congress Series*, 2006, 1293, 253-61.

Gibson, CE. The dynamics of phosphorus in freshwater and marine environments. In: Phosphorus loss from soil to water (H; Tunney, OT; Carton, PC; Brookes, AE. Johnston, Eds), CAB International, London, 1997, 119-135.

Glebe, TW. The environmental impact of European farming: how legitimate are agri-environmental payments? *Review of Agricultural Economics*, 2006, 29, 87-102.

Goldstein, AL; Ritter, GJ. A performance-based non-point source regulatory program for phosphorus control in Florida. In: K. Steele, Editor. Animal Waste and the Land-Water Interface. Lewis Publishers, *Boca Raton*, FL, 1995, 429-41.

Govindasamy, R; Cochran, MJ. Implications of alternative environmental policies on phosphorus loading from poultry litter. *Agricultural Economics*, 1995, 13, 137-48.

Govindasamy, R; Cochran, MJ. Implications of poultry regulations on land applications of poultry litter. *Agricultural and Resource Economics Review*, 1998, 85-94.

Haapapuro, ER; Barnard, ND; Simon, M. Review – animal waste used as livestock feed: dangers to human health. *Preventive Medicine*, 1997, 26, 599-602.

Hamilton, CM; Sims, JT. Nitrogen and phosphorus availability in enriched, pelletized poultry litters. *Journal of Sustainable Agriculture*, 1995, 5(3), 115-32.

Hannawald, J. Alternative Waste Management Technologies: Summary of Available Resources. U.S. Department of Agriculture, Natural Resource Conservation Service. Available at *http:/ /www .nrcs .usda .gov/*

TECHNICAL/ECS/nutrient/manureutil.doc. Last accessed, 1999, October 28, 2009.

Hao, X; Chang, C. Does long-term heavy cattle manure application increase salinity of a clay loam soil in semi-arid southern Alberta? Agriculture, *Ecosystems and Environment*, 2003, 94, 89-103.

Hayes, Adam. Building Soil Organic Matter With Manure. Available at *http://www.omafra.gov.on.ca/english/crops/field/news/croptalk/2003/ct_1 103a2.htm*. *Ontario Ministry of Agriculture*, Food and Rural Affairs. November, 2003.

Hayes, ET; Leek, ABG; Curran, TP; Dodd, VA; Carton, OT; Beattie, VE; et al. The influence of diet crude protein level on odour and ammonia emissions from finishing pig houses. *Bioresource Technology*, 2004, 91, 309-15.

Herath, DP; Weersink, AJ; Carpentier, CL. Spatial and temporal changes in the U.S. hog, dairy and fed-cattle sectors, 1975-2000 *Review of Agricultural Economics*, 2004, 27, 49-69.

Herriges, JA; Secchi, S; Babcock, BA. Living with hogs in Iowa: the impact of livestock facilities on rural residential property values. Working Paper 03-WP 342. August 2003. *Center for Agricultural and Rural Development*, Iowa State University, Ames, IA, 2003.

Honeyman, MS. Environment-friendly swine feed formulation to reduce nitrogen and phosphorous excretion. *American Journal of Alternative Agriculture*, 1993, 8(3), 128-132.

Hooda, PS; Edwards, AC; Anderson, HA; Miller, A. A review of water quality concerns in livestock farming areas. *The Science of the Total Environment*, 2000, 250, 143-67.

Huang, H; Miller, GY. Citizen complaints, regulatory violations, and their implications for swine operations in Illinois. *Review of Agricultural Economics*, 2006, 28, 89-110.

Hubbard, RK; Newton, GL; Hill, GM. Water quality and the grazing animal. *Journal of Animal Science*, 2004, 82 (2004 E Supplement): E255-263.

Huijsmans, J; Verwijs, B; Rodhe, L; Smith, K. Costs of emission-reducing manure application. *Bioresource Technology*, 2004, 93, 11-19.

Innes, R. The economics of livestock waste and its regulation. American *Journal of Agricultural Economics*, 2000, 82, 97-117.

Isik, M. Environmental regulation and the spatial structure of the U.S. dairy sector. *American Journal of Agricultural Economics*, 2004, 86, 949-62.

Johansson, RC; Kaplan, JD. A carrot-and-stick approach to environmental improvement: marrying agri-environmental payments and water quality regulations. *Agricultural and Resource Economics Review*, 2004, 33, 91-104.

Jones, R.; Frarey, L; Bouzaher, A; Johnson, S. Neibergs, S. Livestock and the Environment: A National Pilot Project, Detailed Problem Statement. *Texas Institute for Applied Environmental Research. Stephenville*, Texas. 1993.

Kaplan, JD; Johansson, RC; Peters, M. The manure hits the land: economic and environmental implications when land application of nutrients is constrained. *American Journal of Agricultural Economics*, 2004, 86, 688-700.

Keith, G., Norvell, S; Jones, R; Maguire, C; Osei, E; Saleh, A; Gassman, P. John Rodecap. Livestock and The Environment: A National Pilot Project: CEEOT-LP Modeling for the Upper Maquoketa River Watershed, Iowa: Final Report. Report No. PR0003. Stephenville, Texas: Texas Institute for *Applied Environmental Research*, Tarleton State University. 2000.

Kellogg, RL; Charles, H; Lander, David, C. Moffitt, Noel Gollehon. Manure nutrients relative to the capacity of cropland and pastureland to assimilate nutrients: Spatial and Temporal Trends for the United States. Publication No. nps00-0579. *United States Department of Agriculture*, Natural Resources Conservation Service, Economic Research Service. December 2000.

Keplinger, KO; Hauck, LM. The economics of manure utilization: model and application. *Journal of Agricultural and Resource Economics*, 2006, 31, 414-40.

Keplinger, K; Abraham, J. Project Summary: CEEOT-LP modeling for Duck Creek watershed, Texas. Report No. PR0206. *Texas Institute for Applied Environmental Research*, Tarleton State University, Stephenville, Texas. 2002.

Khan, SJ; Roser, DJ; Davies, CM; Peters, GM; Stuetz, RM; Tucker, R; et al. Chemical contaminants in feedlot wastes: concentrations, effects and attenuation. *Environment International*, 2007, doi: 10.1016/j. envint.2007. 10.007.

Khuder, S; Milz, SA; Bisesi, M; Vincent, R; McNulty, W; Czajkowski, K. Health survey of residents living near farm fields permitted to receive biosolids. *Archives of Environmental and Occupational Health*, 2007, 61, 5-11.

Kuipers, A; Mandersloot, F. Reducing nutrient losses on dairy farms in The Netherlands. *Livestock Pollution Science*, 1999, 61, 139-44.

Lakshminarayan, PG; Bouzaher, A; Neibergs, S; Chen, S; Osei, E; Johnson, SR. Dynamics and Trends in the U.S. Dairy Industry, Livestock Series Report 3. *Center for Agricultural and Rural Development*, Iowa State University, Ames, IA, 1994.

Larney, FJ; Hao, X. A review of composting as a management alternative for beef cattle feedlot manure in southern Alberta, Canada. *Bioresource Technology*, 2007, 98, 3221-7.

Larney, FJ; Yanke, LJ; Miller, JJ; McAllister, TA. Fate of coliform bacteria in composted beef cattle feedlot manure. *Journal of Environmental Quality*, 2003, 32, 1508-15.

Lau, DCW; Wu, MMW. Manure composting as an option for utilization and management of animal waste. *Resources and Conservation*, 1987, 13, 145-56.

Lazarus, WF; Koehler, RG. The economics of applying nutrient-dense livestock waste at low rates. *Review of Agricultural Economics*, 2002, 24, 141-59.

Lazarus, WF; Rudstrom, M. The economics of anaerobic digester operation on a minnesota dairy farm. *Review of Agricultural Economics*, 2007, 29, 349-64.

Lee, LS; Carmosini, N; Sassman, SA; Dion, HM; Sepulveda, MS. Agricultural contributions of antimicrobials and hormones on soil and water quality. *Advances in Agronomy*, 2007, 93, 1-68.

Little, Joanne L; Rodney Bennett, D; Jim, J. Miller. Nutrient and sediment losses under simulated rainfall following manure incorporation by different methods. *Journal of Environmental Quality*, 2005, 34, 1883-1895.

Lory, JA; Massey, RE; Zulovich, JM; Hoehne, JA; Schmidt, AM; Carlson, MS; et al. Feasibility and costs of phosphorus application limits on 39 U.S. swine operations. *Journal of Environmental Quality*, 2004, 33, 1114-23.

Lory, JA; Massey, RE; Zulovich, JM; Hoehne, JA; Schmidt, AM; Carlson, MS; et al. An assessment of nitrogen-based manure application rates on 39 U.S. swine operations. *Journal of Environmental Quality*, 2004, 33, 1106-13.

Lusk, P. Methane Recovery from Animal Manures: A Current Opportunities Casebook, 3rd ed. NREL/SR-25145. National Renewable Energy Laboratory, Golden, CO. *Work performed by Resource Development Associates*, Washington, DC, 1998.

Maguire, RO; Dou, Z; Sims, JT; Brake, J; Joern, BC. Dietary strategies for reduced phosphorus excretion and improved water quality. *Journal of Environmental Quality*, 2005, 34, 2093-103.

Maguire, RO; Sims, JT; Saylor, WW; Turner, BL; Angel, R; Applegate, TJ. Influence of phytase addition to poultry diets on phosphorus forms and solubility in litters and amended soils. *Journal of Environmental Quality*, 2004, 33, 2306-16.

Martinez, SW. Vertical coordination in the pork and broiler industries: implications for pork and chicken products. Agricultural Economic Report No. 777. Food and Rural Economics Division, *Economic Research Service*, U.S. *Department of Agriculture*, 1999.

Massey, R; Krishna, VSR. The effect of manure application costs on a Nebraska feedlot. In the Proceedings of the 1995 *Nebraska Economic and Business Association Annual Meeting*. Wayne, Nebraska, 3 November 1994.

Massey, RE; Applegate, T; Koelsch, R. Economics of land application of laying hen manure. In: *Proceedings of the Multi-state Poultry Feeding and Nutrition Conference*, 22-24. May 2007 [unpaginated CD-ROM]. Indianapolis, IN, 2007.

Matulich, SC; Carman, HF; Carter, HO. Systems analysis of livestock waste management: a study of large-scale dairying. *Western Journal of Agricultural Economics*, 1979, 4, 33-42.

McFarland, AMS; Hauck, LM. Relating agricultural land uses to in-stream stormwater quality. *Journal of Environmental Quality*, 1999, 28, 836-844.

McNitt, J; Jones, R; Osei, E; Hauck, L; Jones, H. Livestock and the Environment: precedents for runoff policy; policy option-CEEOT-LP. Report PR9909. *Texas Institute for Applied Environmental Research*, Tarleton State University, Stephenville, Texas. 1999.

Metcalfe, M. State legislation regulating animal manure management. *Review of Agricultural Economics*, 2000, 22, 519-32.

Milk Market Administrator. The milk market administrator's report. Southwest marketing area. Volume XXXIV, No. 3. March, 2008. Carrollton, Texas. Available at www: http://www.dallasma.com.

Moffitt, LJ; Zilberman, D; Just, RE. A 'putty-clay' approach to aggregation of production/pollution possibilities: an application in dairy waste control. *American Journal of Agricultural Economics*, 1978, 60, 452-9.

Monaco, S; Hatch, DJ; Sacco, D; Bertora, C; Grignani, C. Changes in chemical and biochemical soil properties induced by 11-yr repeated additions of different organic materials in maize-based forage systems. *Soil Biology and Biochemistry*, 2008, 40, 608-15.

Morse, D; Head, HH; Wilcox, CJ; Van Horn, HH; Hissem, CD; Harris, Jr B. Effects of concentration of dietary phosphorus on amount and route of excretion. *Journal of Dairy Science*, 1992, 75, 3039-49.

Morse, D., Head, HH; Wilcox, CJ; Van Horn, HH; Hissem, CD; Harris, B. Jr. Effects of concentration of dietary phosphorus on amount and route of excretion, *Journal of Dairy Science*. November, 1992, 75(11), 3039-3049.

Mueller, DH; Wendt, RC; Daniel, TC. Phosphorus losses as affected by tillage and manure application. *Soil Science Society of America Journal*, 1984, 48, 901-905.

Mullen, JD; Centner, TJ. Impacts of adjusting environmental regulations when enforcement authority is diffuse: confined animal feeding operations and environmental quality. *Review of Agricultural Economics*, 2004, 26, 209-19.

Nehring, R; Barnard, C; Banker, D; Breneman, V. Urban influence on costs of production in the Corn Belt. *American Journal of Agricultural Economics*, 2006, 88, 930-46.

Nicholson, FA; Smith, SR; Alloway, BJ; Carlton-Smith, C; Chambers, BJ. An inventory of heavy metals inputs to agricultural soils in England and Wales. *The Science of the Total Environment*, 2003, 311, 205-19.

NRC. Soil and Water quality: An agenda for agriculture. Board on Agriculture, *Committee on Long-range soil and water conservation*, National Research Council. Washington, DC. 1993.

Ondersteijn, CJM; Beldman, ACG; Daatselaar, CHG; Giesen, GWJ; Huirne, RBM. The Dutch mineral accounting system and the European nitrate directive: implications for N and P management and farm performance. Agriculture, *Ecosystems and Environment*, 2002, 92, 283-96.

Osei, E; Babcock, B. Environmental policy, technology substitution and cross-media transfers, Livestock Series Report 6. Center for Agricultural and Rural Development. *Ames*, IA; 1996.

Osei, E; Hauck, LM; Jones, R; Beran, L; Ogg, C; Alexander, S; et al. Livestock and the environment: lessons from a national pilot project. Project Report PR0801. Texas Institute for Applied *Environmental Research*. Tarleton State University, Stephenville, TX, 2008.

Osei, E; Jones, R; Keis, V. Feasibility of a Dairy Park System. August 2007. *Texas Institute for Applied Environmental Research*, Tarleton State University, Stephenville, TX, 2007.

Osei, E; Lakshminarayan, PG. The determinants of dairy farm location. Livestock Series Report 7. *Center for Agricultural and Rural Development*, Iowa State University, Ames, IA, 1996.

Osei, E; Tanter, A. Economic feasibility of a dairy anaerobic digestion system. Report submitted to the Bosque River Authority. August 2007. *Texas Institute for Applied Environmental Research*, Tarleton State University, Stephenville, TX, 2007.

Osei, E;. Babcock, B. Environmental policy, technology substitution and cross-media transfers, Livestock Series Report 6, *Center for Agricultural and Rural Development*, November 1996.

Osei, E., Gassman, PW; Saleh, A. Livestock and the Environment: Economic and Environmental Modeling Using CEEOT. Report No. PR0002. Stephenville, Texas: *Texas Institute for Applied Environmental Research*, Tarleton State University. 2000a.

Osei, E; Gassman, PW; Hauck, LM; Jones, RD; Beran, LJ; Dyke, PT; Goss, DW; Flowers, JD; McFarland, AMS. Economic costs and environmental benefits of manure incorporation on dairy waste application fields. *Journal of Environmental Management*, 2003a, Vol. 68(1)1-11. May 2003.

Osei, E; Gassman, PW; Hauck, LM; Neitsch, S; Jones, RD; McNitt, J; Jones, H. Economic and environmental impacts of pasture nutrient management. *Journal of Range Management*, 2003b, 56, 218-226 May 2003.

Osei, E; Gassman, PW; Jones, RD; Pratt, SJ; Hauck, LM; Beran, LJ; Rosenthal, WD; Williams, JR. Economic and environmental impacts of alternative practices on dairy farms in an agricultural watershed. *Journal of Soil and Water Conservation*, 2000b, 55(4), 466-472.

Osei, Edward, Bing Du, Asfaw Bekele, Larry, M. Hauck, Ali Saleh, and Alex Tanter. 2008. Impacts of alternative manure application rates on Texas animal feeding operations: a macro level analysis. *Journal of the American Water Resources Association*. Forthcoming (June 2008).

Palmquist, RB; Roka, FM; Vukina, T. Hog operations, environmental effects, and residential property values. *Land Economics*, 1997, 73, 114-24.

PATS. 2001. Technology Adoption among Wisconsin's Dairy Farmers in the 1990s. *Program on Agricultural Technology Studies*. Fact Sheet 15. Wisconsin Family Farm Facts.No. 15, June, 2001.

Paudel, KP; Adhikari, M; Martin, Jr NR. Evaluation of broiler litter transportation in northern Alabama, USA. *Journal of Environmental Management*, 2004, 73, 15-23.

Perry, J; Banker, D; Green, R. Broiler farms' organization management, and performance. AIB-748. United States Department of Agriculture, Economic Research Service, *Resource Economics Division*, 1999.

Petersen, SO; Sommer, SG; Be´ line, F; Burton, C; Dach, J; Dourmad, JY; et al. Recycling of livestock manure in a wholefarm perspective. *Livestock Science*, 2007, 112, 180-91.

Pizarro, C; Mulbry, W; Blersch, D; Kangas, P. An economic assessment of algal turf scrubber technology for treatment of dairy manure effluent. *Ecological Engineering*, 2006, 26, 321-7.

Pope III, CA; McBryde, GL. Optimal stocking of rangeland for livestock production within a dynamic framework. *Western Journal of Agricultural Economics*, 1984, 9, 160-9.

Powell, JM; Fernandez-Rivera, S; Hofs, S. Effects of sheep diet on nutrient cycling in mixed farming systems of semi-arid West Africa. Agriculture, *Ecosystems and Environment*, 1994, 48, 263-71.

PPI, Potash and Phosphate Institute. 2001. Soil test levels in North America: summary update. PPI/PPIC/FAR Technical Bulletin 2001-1. *Potash and Phosphate Institute*, Norcross, Georgia, U.S.A. 2001.

PPI, Potash and Phosphate Institute. 2002. Plant nutrient use in North American agriculture. PPI/PPIC/FAR Technical Bulletin 2002-1. *Potash and Phosphate Institute*, Norcross, Georgia, U.S.A. 2002.

Pratt, Staci, Jones, Ron and Jones, C. Allan. Livestock and the environment: Expanding the focus: Policy options: CEEOT-LP. Report PR9603. *Texas Institute for Applied Environmental Research*, Tarleton State University, Stephenville, Texas. 1997.

Progressive Farmer. 1998. Phytase costs can be a wash. Progressive Farmer. http://progressivefarmer.com/issue/0798/pork/default.asp July 1998. Site visited July 28, 1999.

Radcliffe, JS; Kornegay, ET. Zhang, Z. Effectiveness of microbial phytase as a feeding strategy for reducing phosphorus excretion in pigs and poultry. In Animal Production Systems and the Environment: An International Conference of Odor, Water Quality, *Nutrient Management,* and Socioeconomic Issues: Proceedings Vol. I. Ames, IA: Iowa State University. 1998.

Ready, R; Abdalla, C. GIS analysis of land use on the rural–urban fringe: the impact of land use and potential local disamenities on residential property *values and on the location of residential development in Berks County*, Pennsylvania. Staff paper 364. Pennsylvania State University, University Park, PA, 2003.

Ribaudo, M; Agapoff, J. Importance of cost offsets for dairy farms meeting a nutrient application standard. *Agricultural and Resource Economics Review*, 2005, 34, 173-84.

Ribaudo, M; Cattaneo, A; Agapoff, J. Cost of meeting manure nutrient application standards in hog production: the roles of EQIP and fertilizer offsets. *Review of Agricultural Economics*, 2004, 26, 430-44.

Ribaudo, M; Gollehon, N; Aillery, M; Kaplan, J; Johansson, R; Agapoff, J; et al. Manure Management for Water Quality: Costs to Animal Feeding Operations of Applying Manure Nutrients to Land. Agricultural Economic Report 824. U.S. Department of Agriculture, *Economic Research Service, Resource Economics Division*, 2003.

Roka, FM; Hoag, DL. Manure value and liveweight swine decisions. *Journal of Agricultural and Applied Economics*, 1996, 28, 193-202.

Romkens, MJM; Nelson, DW; Mannering, JV. Nitrogen and phosphorus composition of surface runoff as affected by tillage method. *Journal of Environmental Quality*, 1973, 2 (2), 292-295.

Rotz, CA; Sharpley, AN; Satter, LD; Gburek, WJ; Sanderson, MA. Production and feeding strategies for phosphorus management on dairy farms. J. *Dairy Sci.*, 2002, 85, 3142-3153.

Satter, Larry, D; Zhiguo Wu. "New strategies in ruminant nutrition: getting ready for the next millennium," *in Southwest Nutrition & Management Conference Proceedings*, February 25-26, 1999 (The University of Arizona, Tucson, AZ). 1999.

Schiffman, SS; Sattely Miller, EA; Suggs, MS; Graham, BG. The effect of environmental odors emanating from commercial swine operations on the mood of nearby residents. *Brain Research Bulletin*, 1995, 37, 369-75.

Schjonning, P; Elmholt, S; Munkholm, LJ; Debosz, K. Soil quality aspects of humid sandy loams as influenced by organic and conventional long-term management. Agriculture, *Ecosystems and Environment*, 2002, 88, 195-214.

Schmit, TM; Knoblauch, WA. The impact of nutrient loading restrictions on dairy farm profitability. *Journal of Dairy Science*, 1995, 78, 1267-81.

Schnitkey, GD; Miranda, MJ. The impact of pollution controls on livestock-crop producers. *Journal of Agricultural and Resource Economics*, 1993, 18, 25-36.

Sharpley, AN; Chapra, SC; Wedepohl, R; Sims, JT; Daniel, TC; Reddy, KR. Managing agricultural phosphorus for protection of surface waters: issues and options, *Journal of Environmental Quality*, 1994, 23, 437-51.

Sharpley, AN; Rekolainen, S. Phosphorus in agriculture and its environmental implications. In: H; Tunney, OT; Carton, PC; Brookes, AE. Johnston, Editors. *Phosphorus Loss from Soil to Water*. CAB International, Wallingford, UK, 1997. 1-54.

Sharpley, AN; Rekolainen, S. Phosphorus in agriculture and its environmental implications. In: Phosphorus loss from soil to water (H; Tunney, OT; Carton, PC; Brookes, AE. Johnston, Eds), CAB International, London, 1997, 1-53.

Sharpley, Andrew N; Chapra, SC; Wedepohl, R; Sims, JT; Daniel, TC; Reddy, KR. Managing agricultural phosphorus for protection of surface waters: issues and options, *Journal of Environmental Quality*, 1994, 23, 437-451.

Shepard, Robin. Nitrogen and phosphorus management on Wisconsin farms: Lessons learned for agricultural water quality programs. *Journal of Soil and Water Conservation*, 2000, 55(1), 63-68.

Sheriff, G. Efficient waste? Why farmers over-apply nutrients and the implications for policy design. *Review of Agricultural Economics*, 2005, 27, 542-57.

Sheriff, G. Efficient waste? Why farmers over-apply nutrients and the implications for policy design. *Review of Agricultural Economics*, 2005, 27, 542-57.

Sonneveld, MPW; Schroder, JJ; de Vos, JA; Monteny, GJ; Mosquera, J; Hol, JMG; et al. A whole-farm strategy to reduce environmental impacts of nitrogen. *Journal of Environmental Quality*, 2008, 37, 186-95.

Sutton, AL; Jones, DJ; Joern, B. Total Farm Nutrient Management – Manure Treatment. Purdue Animal Issues Briefing. AI-9. 2/99. Purdue University. Available at *http://www.ansc.purdue.edu/anissue/AI9.pdf*. Last accessed, 1999. October 28, 2009.

Sweeten, JM; Miner, JR. Odor intensities at cattle feedlots in nuisance litigation. *Bioresource Technology*, 1993, 45, 177-88.

Sweeten, John. Personal communication. Director, *Texas Agricultural Experimetn Station*, Amarillo, Texas. April, 1998.

Syers, JK; Harris, RF; Armstrong, DE. Phosphate chemistry in lake sediments. *Journal of Environmental Quality*, 1973, 2, 1-14.

Tweeten, L. Agricultural Industrialization: For Better or Worse? Anderson Chair Occasional Paper ESO #2404. *Department of Agricultural, Environmental, and Development Economics*, The Ohio State University; 1998.

USDA NASS. U. S. Census of Agriculture: 1982. National Agricultural Statistics Service. *US Department of Agriculture*. Washington DC. 1982.

USDA NASS. U. S. Census of Agriculture: 1992. National Agricultural Statistics Service. *US Department of Agriculture*. Washington DC. 1992.

USDA NASS. U. S. Census of Agriculture: 2002. National Agricultural Statistics Service. *US Department of Agriculture*. Washington DC. 2002.

USDA NRCS. Comprehensive nutrient management planning technical guidance. United States Department of Agriculture, *Natural Resources Conservation Service*, 2000, 1 December.

USDA. Costs associated with development and implementation of comprehensive nutrient management plans: Part I—Nutrient management, land treatment, *manure and wastewater handling and storage*, and recordkeeping. USDA NRCS. June 2003.

USDA. Costs associated with development and implementation of comprehensive nutrient management plans: part I – nutrient management, land treatment, manure and wastewater handling and storage, and recordkeeping. United States Department of Agriculture, *Natural Resources Conservation Service*, June 2003.

USEPA. National water quality inventory 2000 Report to Congress. U.S. Environmental Protection Agency. Office of Water. U.S. *Government Printing Office*, Washington, DC. 2002.

USEPA. AgStar Handbook: *A Manual for Developing Biogas Systems at Commercial Farms in the United States*. EPA 430-B-97-015. Washington, DC, 1997.

USEPA. Inventory of U.S. greenhouse gas emissions and sinks: 1990-2005. U.S. *Environmental Protection Agency*. Office of Atmospheric Programs. Available at http:// www. epa. gov/ climatechange /emissions/ usinventoryreport. html. Report No. EPA 430-R-07-002. Washington DC, April 15, 2007.

USEPA. Alternative technologies/uses for manure. Undated. Available at *http://www.epa.gov/npdes/pubs/cafo_report.pdf*. Last accessed, 2009, October 28.

USEPA. Inventory of U.S. Greenhouse Gas Emissions and Sinks: 1990-2003. U.S. *Environmental Protection Agency*. Washington DC, 15 April 2005. EPA 430-R-05-003, 2005.

USEPA. Inventory of U.S. Greenhouse Gas Emissions and Sinks: 1990-2005. U.S. *Environmental Protection Agency*. Office of Atmospheric Programs. Report No. EPA 430-R-07- 002. Washington DC, 15 April 2007. Available from: URL: http://www.epa.gov/climatechange/emissions/ usinventoryreport.html.

USEPA. United States Environmental Protection Agency. 2002 National Water Quality Inventory. EPA-841-R-02-001. Available from: URL: http://www.epa.gov/305b/2000report/ (accessed 8 July 2003).

Van Horn, HH; Newton, GL; Kunkle, WE. Ruminant nutrition from an environmental perspective: factors affecting whole-farm nutrient balance. *Journal of Animal Science*, 1996, 74, 3082-102.

Van Horn, HH. Recycling manure nutrients to avoid environmental pollution. *In Large Dairy Herd Management*, HH; Van Horn, CJ. Wilcox, Eds. (Champaign, IL: American Dairy Science Association, 1992).

Van Horn, HH; Newton, GL; Kunkle, WE. Ruminant nutrition from an environmental perspective: factors affecting whole-farm nutrient balance, *Journal of Animal Science*, 1996, 74, 3082-3102.

Van Horn, HH. Achieving environmental balance of nutrient flow through animal production systems. *The Professional Animal Scientist*, 1991, 7(3), 22-33.

Vázquez, FJ; Petrikova, V; Villar, MC; Carballas. T. Use of poultry manure and plant cultivation for the reclamation of burnt soils. *Biology and Fertility of Soils*. Volume 22, Number 3 / May, 1996.

Vázquez1, FJ; Petrikova1, V. 2, M. C. Villar1 and T. Carballas1 Vergura, Jim, Ron Jones, Staci Pratt, Gary Keith. 2002. PIMA: An alternative approach to watershed management. Report PR0203. *Texas Institute for Applied Environmental Research*, Tarleton State University, Stephenville, Texas. March, 2002.

Vervoort, RW; Keeler, AG. The economics of land application of fresh and composted broiler litter with an environmental constraint. *Journal of Environmental Management*, 1999, 55, 265-72.

Vogt, ML; Kastens, TL. A study of the financial impact of dairy manure storage systems in northeast Kansas. *Review of Agricultural Economics*, 2005, 27, 336-49.

Vukina, T; Wossink, A. Environmental policies and agricultural land values: evidence from the Dutch nutrient quota system. *Land Economics*, 2000, 76, 413-29.

Vukina, T. The relationship between contracting and livestock waste pollution. *Review of Agricultural Economics*, 2003, 25, 66-88.

Warrick, J; Leavenworth, S. Who will rescue the ravaged river? *The News and Observer*. Raleigh, NC, March, 1999.

Webb, J; Menzi, H; Pain, BF; Misselbrook, TF; Dammgen, U; Hendriks, H. Managing ammonia emissions from livestock production in Europe. *Environmental Pollution*, 2005, 135, 399-406.

Webster. 2009."Manure". In Merriam-Webster Online Dictionary. Retrieved October 28, 2009, from http:// www. merriam-webster .com /dictionary /manure.

Weersink, A; deVos, G; Stonehouse, P. Farm return and land price effects from environmental standards and stocking density restrictions. *Agricultural and Resource Economics Review*, 2004, 33, 272-81.

Williams, JR; Arnold, JG; Srinivasan, R. The APEX Model. BRC Report No. 00-06, Texas A&M. *Blackland Research and Extension Center*, Temple. 2000.

Willinger, H. Odour and pathogen control from intensive animal and poultry husbandry in Austria. *Agriculture and Environment,* 1974, 1, 39-50.

Wimbely, J; Goodwin, HL. Alternative Poultry Litter Management in the Eucha/Spavinaw Watershed. Part I: Raw Litter Export: *A Report to the Tulsa Metropolitan Utility Authority*. Tulsa, OK, 2000.

Wing, S; Wolf, S. Intensive livestock operations, health, and quality of life among eastern North Carolina residents. *Environmental Health Perspectives*, 2000, 108, 233-8.

Wossink, A; Gardenbroek, C. Environmental policy uncertainty and marketable permit systems: the Dutch phosphate quota program. *American Journal of Agricultural Economics*, 2006, 88, 16-27.

Wu, Z; Satter, LD. Milk production and reproductive performance of dairy cows fed two concentrations of phosphorus for two years. *Journal of Dairy Science*, 2000, 83, 1052-63.

Wu, Z; Satter, LD. Milk production and reproductive performance of dairy cows fed two concentrations of phosphorus for two years. *J Dairy Sci*., 2000, 83, 1052-1063.

Xu, F; Prato, T. Optimal farm-level use and value of broiler litter. In: Steele K, editor. Animal Waste and the Land-Water Interface. Lewis Publishers, *Boca Raton*, FL, 1995, 283-91.

Yap, C; Foster, K; Preckel, P; Doering, O; Richert, B. Mitigating the compliance cost of a phosphorus-based swine manure management policy. *Journal of Agricultural and Applied Economics*, 2004, 36, 23-34.

Zhao, Suling, L., Satish, C. Gupta, David, R. Huggins and John F. Moncrief. 2001. Tillage and nutrient source effects on surface and subsurface water quality at corn planting. *Journal of Environmental Quality*, 30, 998-1008.

In: Manure: Management, Uses and… ISBN: 978-1-61668-424-2
Editors: C. S. Delliaguardia, pp. 49-85

Chapter 2

Manure: Its Impacts on Environment Quality and on Soil Biota with Special Reference to Nematodes

Qudsia Tahseen*

Aligarh Muslim University, Aligarh, India

Abstract

Soil health can be boosted by improving soil texture, aeration, drainage, and nutritional content by application of manure/amendments besides promoting the growth of the beneficial biota and suppressing the pathogens. Manures can vary in their nutrient contents and forms, depending on their processing and storage procedures, feeds and feed supplements used, etc. Of the total N, 30% to 90% is comprised of ammonium while the rest is contained in organic forms that must be decomposed (mineralized) to inorganic forms to be available to plants. Likewise, the phosphorus in manure tends to be fixed in soils by sorption and precipitation reactions and its mineralization takes place to make it available to plants. Manures are also effective sources of potassium (K) and sulphur (S). The manure-enriched soils enhance the populations of mites and collembolans besides increasing the number of fungi that directly affect the population of soil nematodes. Some of the fungi use adhesive conidia, branches, knobs and mycelia to parasitize nematodes.

---

* Nematode Research Laboratory, Department of Zoology, Aligarh Muslim University, Aligarh-202002, India, E-mail: qtahseen@yahoo.com

However, compared to plant parasitic nematodes, the density of opportunistic bacterial-feeding and fungal-feeding nematodes increases significantly in the soils treated with manure. Among the manures of animal origin poultry and farmyard manures give significantly higher yields than those of other organic manures. The impacts of manures also vary considerably in the soil systems. Poultry manure causes short term (within a growing season) increases in microbiological activity and nutrient cycling, whereas farm yard manure seems to bring long-term changes. Green manure application too has gained wide acceptance due to the improvement in soil characteristics, including drainage and water retention, nutrient content and storage, and level of microbiological activity in the soil.

I. Introduction

Soil is one of the most poorly researched habitats on earth. The existence of this thin dark covering on the surface of the earth is vital for the survival of the biosphere in its present form. The soil biota as a result are responsive to various induced disturbances like agricultural practices, deforestation, pollution, disturbance in global elemental cycle, erosion, green house gas influx etc.

Soil is composed of the non-living component with various minerals and compounds determining its composition, and the living part reflecting the organisms associated. There exists an intricate relationship between these two components that ensures soil health. Soil organic matter is any material in the soil that is originally produced by soil biota. It may comprise of a variety of materials from the intact tissues/ by products of organisms to the substantially decomposed ones. The original plant/ animal tissues are formed of different organic compounds which decompose later at different rates. The complex interactions between the organic waste, microorganisms and other soil fauna animals result in the biooxidation and stabilization of wastes. A large variety of microorganisms and soil invertebrates proliferate and interact contributing to the "cycle of matter". Billions of soil bacteria of different kinds recycle organic matter and help in C: N mineralization. One of the most important groups in that regard is of nitrogen-fixing bacteria. They consume particles of organic matter, creating a waste product grossly called bacterial manure thus adding new forms of organic content to soil. Most plants absorb nutrients most efficiently through this bacterial waste product thus showing a better growth in soils with greater bacterial biomass. Like bacteria, fungi break down organic matter by digesting and excreting it, thus recycling nutrients through the soil food web. Vesicular Arbuscular Mycorrhizae (VAM) are among the best known fungi that grow around the roots of plants thus creating a mesh of fine feeder "rootlets" that facilitate absorption of nutrients and water by the plant root system. In effect, they increase the surface area of the roots. The other subterranean fauna playing significant role in matter recycling include the micro arthropods (mites, sow bugs, springtails, spiders, and centipedes) and the nematodes that consume debris and bacteria to make nitrogen and other nutrients more readily available to plants and other soil biota. The protozoans serving as food for most soil predators, release nitrogen and other nutrients back into the soil through excretion. The earthworms are very beneficial as they tunnel through heavy soil to let air get down to plant roots and also turn the soil layers thus acting as mechanical blenders to break the soil crumbs through their movement, besides contributing to recycling of matter. As a result of decomposition by bacteria, fungi, and actinomycetes, the most readily available energy sources (carbohydrates, fats, proteins) are used up by the microorganisms leaving behind a dark mixture usually referred to as 'humus'. Newly-formed humus is a combination of resistant materials from the original plant tissue and compounds synthesized as part of the microorganisms' tissue that die during the process. The resulting humus is relatively resistant to microbial degradation; hence its nitrogen and other essential nutrients are protected from ready solubility and dissipation. Humus holds water and minerals extremely well and helps soil establish and maintain a strong crumb structure. It provides some nutrients as it is very slowly decayed by microbial activity. Thus soil biota along with non-living components, play an important role in regulating the soil processes. Soil health, therefore, can be boosted by promoting the growth of the beneficial biota and suppressing the pathogens besides improving soil texture, aeration, drainage, and nutritional content by application of amendments.

The use of organic and inorganic amendments to soil has been in practice for centuries to improve soil fertility and increase crop yield. The various forms of organic matter used were compost (organic material decomposing under considerably controlled conditions leading to the exhaustion of readily-exploited nutrients), undecomposed plant and animal materials (leaves and garden waste or harvest residues e.g., fresh pine straw, sawdust etc.) and green manure (fresh plant material grown to be incorporated into the soil where it was grown).

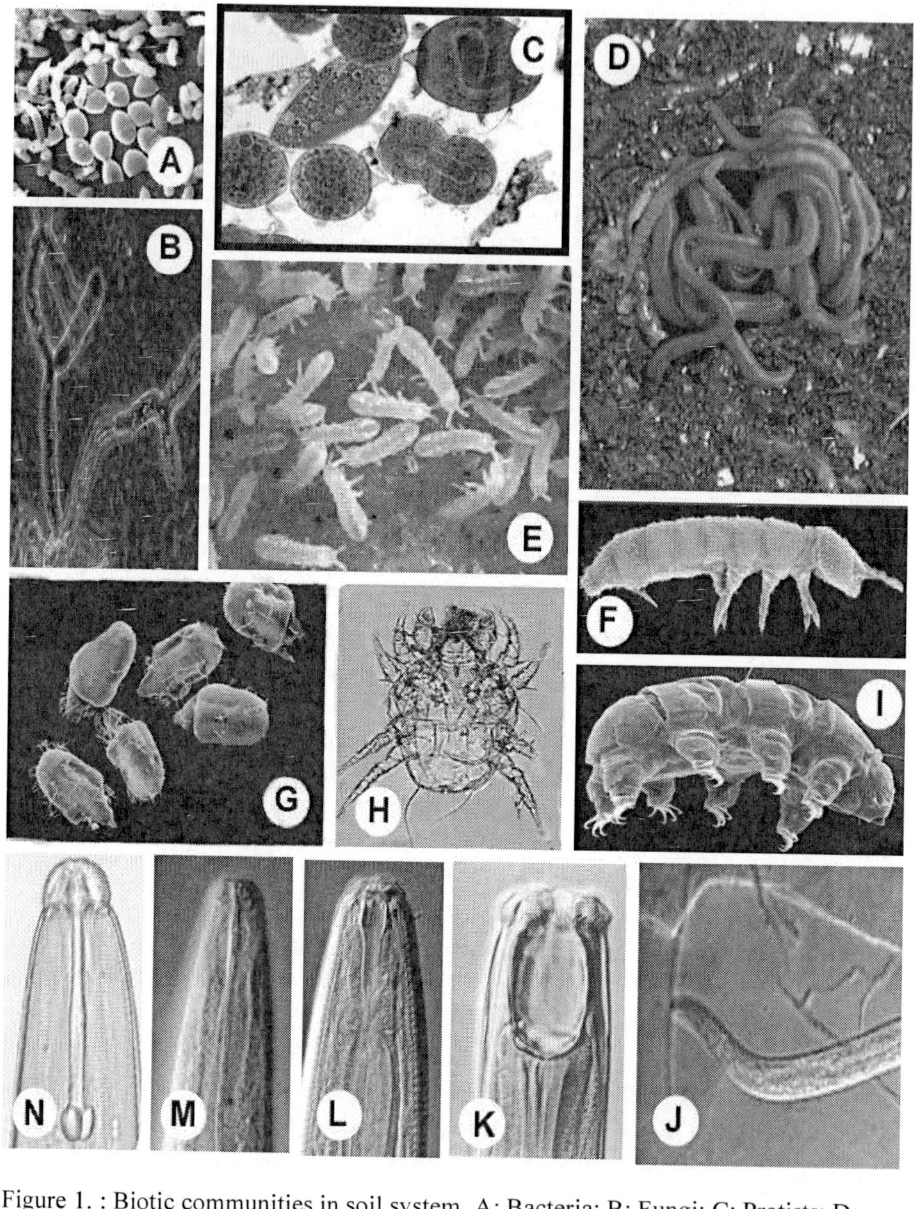

Figure 1. : Biotic communities in soil system. A: Bacteria; B: Fungi; C: Protists; D: Earthworms; E: Springtails/ collembolans; F: *Isotoma* sp.; G, H: Mites; I: Tardigrade; J: Nematode trapped by fungus; K: Predatory nematode; L: Bacteriophagous nematode; M: Mycophagous nematode; N: Plant parasitic nematode.

II. Soil Food Web

Despite the fact that the belowground diversity usually exceeds that which prevails above ground, it has not been fully documented in any system (Anderson, 1995; Freckman *et al.*, 1997). From virtually any perspective, the soil biota is exceptionally diverse. In addition to plants, soil is the habitat of a diverse array of organisms: archeae, bacteria, fungi, protists, algae, nematodes, tardigrades, earthworms, collembolans, mites and many other types of arthropods (Fig.1). There exist many food chains that are woven together into complex food web. At every trophic interaction, the debris and leakage become substrate for other organisms; the wastes and secretions of the herbivores and their carbon molecules, originally fixed by the autotrophs, are substrate for other trophic groups. Carbon, nitrogen, and other molecules are mineralized during the metabolic processes of all organisms in the web through respiration, osmoregulation etc. The mineralized molecules (CO_2, NH_4, NO_3, etc.) are available to plants, bacteria and fungi in the soil, or are returned to the atmosphere. Micro-organisms in the soil environment tend to be aggregated in areas where carbon and energy originally enters the food web, e.g., in the plant rhizosphere, close to the soil surface, or in the tillage zone. The size of the food web is determined by the amount of carbon and energy entering it as about 90% of the remaining resources may be lost at each trophic interchange. That limits the length of chains or channels running through the web to four or five interchanges.

III. Nematodes in Subterranean Food Web

Nematodes are found in sediments or subterranean environment thus constituting an important part of the soil fauna. They live in the maze of pore canals and pores existing between soil particles. Their movement is facilitated by the film of water present in soil pores. Nematodes are important participants in underground energy-transfer system as they regulate mineralization processes along with protists. They consume living plant material, fungi, bacteria, small protists and even each other. Different species of nematodes have specific soil and climatic requirements. For example, certain species do best in sandy soils, while others prefer clayey soils. The populations of plant parasitic nematodes are generally greater and more prevalent in the tropical regions due to longer growing seasons and increased reproductive rates (Dropkin, 1980). These nematodes are found in associated with most plants. Some are endoparasitic, living and feeding within the tissue of roots, tubers, buds, seeds, etc. (Sasser, 1990) while others are ectoparasitic, feeding externally on epidermal and cortical tissues. Barring few host-specific species, such as *Heterodera glycines* on soybeans and *Globodera rostochiensis* on potatoes (Sasser, 1990), the nematodes have a wide host range. Direct feeding by nematodes can drastically decrease a plant's vigour and its potential for uptake of nutrients and water due to impaired root system. They usually prefer light, sandy soils over clayey soils mainly due to efficient aeration and the ease with which they can move through the root zone. However, the plants growing in readily drainable soils are more likely to suffer from intermittent drought, and are thus more vulnerable to damage by these parasites.

Usually nematodes are regarded as silent enemies, however, losses up to 80% have been found in vegetables (Siddiqi, 2000; Kaskavalci, 2007). The problem has commonly been solved by chemical treatment and crop rotation (Murphy *et al.*, 1974). Despite the advantages, both practices have their own limitations. Crop rotation though reduces the population of specific nematode parasites may increase other species to damaging levels (Daulton, 1963; Brodie *et al.*, 1970; Minton and Donnelly, 1971) while costly chemicals/ nematicides cause environmental hazards due to residual impact.

Chemical treatment of soil though generally improves crop yields rarely reduces nematode population densities for more than 2-3 months, resulting in post harvest population densities growing up more than pre plant densities (Brodie and Good, 1973). One of the alternative strategies for management of plant parasitic nematodes is, therefore, the application of organic manure/ amendments (Agyarko and Asante, 2005) that have beneficial impacts on soil environment (Oka *et al.* 2000). There is substantial evidence that the addition of organic matter in the form of compost or manure will decrease nematode pest populations and associated damage to crops (Stirling, 1991; Akhtar and Alam, 1993; Oka and Yermiyahu, 2002; Walker, 2004). This could be a result of improved soil structure and fertility, enhanced level of plant resistance, release of nemato-toxins, or increased populations of fungal and bacterial parasites/ pathogens and other nematode-antagonistic agents (Akhtar and Malik, 2000). Reduced nematode damage from increased organic matter in soil is likely a combination of these interactions. Higher organic matter content increases soil's water-holding capacity, and supports thriving communities of the decomposers and predators including saprophagous and predatory nematodes resulting in improved soil health.

IV. Types of Manures Used in Agriculture

The sources of organic matter known to be nematode-suppressive are animal- or plant-derived including bone meal, horn meal, compost, and certain green manures.

A. Manures of Animal Origin

The manures may be derived from the excreta of animals having rich nitrogenous content. The other types may include the processed manures that are produced by the earthworm culture or by treatment of sewage.

i) Pig Manure

Pig manure is important in the pig-vegetable-fish-duck chain that is commonly used in Southeast Asia largely by Chinese farmers. Swine manure contains over 20% crude protein.

ii) Cattle Manure

Low digestible energy content and relatively low crude protein content are the two primary deficiencies of cattle excreta.

iii) Poultry Manure

With the intensification of poultry, the poultry manure has high applicability in agriculture and animal feed programmes. Fresh poultry manure is about 30% crude protein on a dry basis, about half of which derives from uric acid. Poultry manure is also rich in minerals.

iv) Guano

Droppings from birds or bats contain up to 10% nitrogen and are rich in phosphorus (up to 5% for fish-eating birds). It should be used in the same way as urea and usually gives the same response.

v) Vermicompost

It is produced as a result of complex interactions between the organic waste, microorganisms, earthworms and other soil invertebrates that lead to biooxidation and stabilization of wastes. Earthworms consume and incorporate into soil large amounts of organic matter (MacKay and Kladivko, 1985;

Parmelee *et al.*, 1990) and play an important role in the carbon cycle and the mineralization of nutrients from organic matter in many ecosystems (Hendrix *et al.*, 1987; Edwards *et al.*, 1995). Thus vermicomposting system sustains complex food webs besides modifying different chemical forms of nutrient elements into stable organic compounds, which are important for nutrient dynamics (Domínguez *et al.,* 1997) in the soil environment. The microbes are responsible for the biochemical degradation of organic matter in the vermicomposting process while earthworms help in conditioning the substrate and promoting microbial activity by their castings.

vi) Chitinous Material

The crushed shells of crustaceans (shrimp, crab, etc.) have been found to be effective soil amendments and a good source of organic matter.

vii) Sewage

Activated sewage sludge is made by a fermentation process, in which the organic matter in sewage is metabolized by a variety of microorganisms under heavy aeration. This process is most common in modern sewage treatment plants for the production of large quantity of activated sludge. This end product is a gelatinous mass of microorganisms, rich in nitrogen and vitamin B12.

B. Manures of Plant Origin

The various plant materials used in manures include press-mud, vegetable- and fruit-processing waste and tobacco wastes. However, amendments to soil with spent tea, wheat straw, paddy husk, paddy straw, oilcakes, sawdust, sugarcane bagasse, sugarcane trash, domestic garbage, dead vegetation and pigeon pea stubble are also found to be effective in improving the soil health. Plant oil cakes are also used to improve the yield of various crops. Neem cakes, made from crushed neem seeds, provide nitrogen in a slow-release form in addition to protecting plants against parasitic nematodes.

i) Compost

It is a type of manure that largely utilises the degraded plant material mixed with by-products of degrading organisms. Its preparation is called composting. Substrate size, temperature as well as the nature of material and the dimension of the area, are the important factors determining the rate of composting.

ii) Green Manure

Green manuring is the practice of growing plants in the fields where the incorporation of organic matter is desired. The plant material used in this way is called a green manure. The plant material may or may not be cut free from its roots before being incorporated into the soil, depending on what is needed to be able to handle it. Green manuring is popular among "pure" organic farmers and is widely adopted due to the fact that adding large amounts of green plant material may prove beneficial for soil characteristics, including drainage and water retention, nutrient content and storage, and level of microbiological activity in the soil. Some of the green manure crops include: lespedeza, common vetch, velvet bean, lupin, castor, rapeseed, giselba mustard, and marigold. Urea though equally as effective a nematicide, does not provide the added benefits of green manuring. Green manures, besides adding organic matter to the soil to improve its physical properties such as tilth and water holding capacity, also improve the soil's nutrient status by adding N, P, K, and other nutrients. The plants e.g., castor bean, chrysanthemum, partridge pea, velvet bean, sesame, jack bean, crotalaria, sorghum-sudan, indigo etc. produce allelochemicals that function as antagonistic compounds. Polythienyls, glucosinolates, cyanogenic glycosides, alkaloids, lipids, terpenoids, steroids, triterpenoids, and phenolics are some of the compounds that are exuded during the growing season or released during green manure decomposition. Sunn hemp (tropical legume), and sorghum-sudan produce the allelochemicals known as monocrotaline and dhurrin, respectively. (Chitwood, 2002; Grossman, 1988; Hackney and Dickerson, 1975; Quarles, 1993; Wang *et al.*, 2002; Williams and Williams, 1990a, 1990b, 1993). The products of the neem tree *(Azadirachta indica*) have been found to enrich the soil when used as a soil amendment, and to control plant-parasitic nematodes. It is sold as a fertilizer in many farm and garden supply stores. Neem cakes can be mixed with fertilizers such as composted manures, seaweed, and kelp.

V. General Composition of Manure

The types of nutrient and their contents in different manures may be variable and the laboratory analysis of a sample of specific manure can give the type and value of the nutrients and will be useful in determining the appropriate rate of application for the crop to be grown. Because of the high variability in nutrient content even within a single storage unit, the combined analysis of several sub-samples is required to ascertain average nutrient levels.

i. Nitrogen

Considerable variability can exist in manure nutrient content and form, depending on type of livestock operation, handling and storage procedures, feeds and feed supplements used, etc. Of the total N, 30% to 90% is comprised of ammonium while the rest is contained in organic forms that must be decomposed (mineralized) to inorganic forms to be available to plants. Owing to differences in the forms of N present in manure, different patterns in availability are observed over time. Composting or decomposition of manure in the soil gradually increases the N content of the manure relative to the carbon content over time and allows N to be eventually released. For this reason, solid manures with straw bedding are usually more slowly available sources that may take several years to release their N.

ii. Phosphorus

The total phosphorus (P) content is also directly related to the solid content of the manure and tends to be more variable than the N content. In liquid manure 10% to 50% of the total P is present as readily soluble inorganic phosphate. Of the solid manures, poultry manure is highest in P. Manure P tends to be fixed in soils by sorption and precipitation reactions and the mineralization of organic P takes place to make it available to plants.

iii. Potassium

Manures are effective sources of potassium (K) for plant growth, as the K in manure is readily available.

iv. Sulphur

Some animal manures like, liquid swine effluent tend to be low in sulphur (S) relative to N.

v. Micronutrients

Manures also contain micronutrients including copper, manganese, zinc and boron. There is less information on the forms and availability of micronutrients in manures than macronutrients, as their chemistry in manures and soils is complex and intricate. Micronutrient metals may be present in manures as soluble free and complex cations, as well as adsorbed and precipitated in the solid phase.

VI. Impact of Manure on Subterranean Communities

A. Effect on Earthworms

Higher populations and biomass production in *Lumbricus* sp. were observed in corn plots side dressed with solid dairy cattle manure. The cocoon production increased significantly in manure treatments indicating the potential effects of manure side-dressing on earthworm population dynamics (Coderre *et al.,* 1995). The increase in microbial biomass in the manure treatment was probably stimulated by the enhanced incorporation of manure into the soil by earthworms (Permelee & Crossley, 1988). Cow manure contains readily mineralizable substrates that stimulate microbial growth (Coderre *et al.,* 1995). Studies have shown that *L. terrestris* increases the transfer of nutrients from organic matter on the soil surface to plant tissues (Binet and Trehen, 1992; Devliegher and Verstraete, 1997). Furthermore, this effect of *L. terrestris* earthworms on nutrient transfer from surface residues to plants appears to be greater for plant residues with a wide C/N ratio than for those with a low C/N ratio (Hameed *et al.*, 1993). Addition of farmyard manure increases earthworm populations under favourable soil conditions. However, the cattle slurry and urine may be toxic to earthworms due to the ammonia, benzoic acid and sodium sulphide contents (Curry, 1976). Aeration of slurry has been proposed as a strategy for decreasing the toxicity of the slurry to earthworms.

B. Effect on Mites, Collembolans Etc.

The manure enriched soils enhance the populations of mites and collembolans (Axelsen and Kristensen, 2000), however, higher doses of cattle liquid manure seem to decrease the populations of mites (Graczyk *et al.*, 2009). Some other manure-inhabiting species of mesostigmatid mites i.e., *Macrocheles muscaedomesticae*, *Parasitus* sp., and *Fuscuropoda* sp. were also reported to kill free-living nematodes (Yasuhiro, 1981). Other studies indicated that urea fertilizers showed no persistent effect on mite assemblages, whereas composted chicken manure supported their abundant and diverse populations (Minor and Norton, 2004). The enchytraeids react to liquid manure more distinctly than the mites. Small and medium doses of this fertilizer stimulate their density while the highest doses decrease it. Contrary to it a low dose of liquid manure slightly decreases the density of springtails (collembolans), while higher doses successively increase it (Domek-Chruoecicka and Seniczak, 2005)

C. Effect on Fungi

Experiments have revealed that nematode-destroying fungi increase in number when organic substrates are incorporated into the soil (Gomes *et al.*, 2000; Timm *et al.*, 2001). Organic amendments especially chicken manure stimulated build-up of nematode-destroying fungi, *Arthrobotrys oligospora*, *A. superba*, *A. dactyloides*, *Harposporium anguillulae* and *Monacrosporium cionopagum* (Wachira *et al.*, 2009). Mahoney and Strongman (1994) reported greatest number of species and highest prevalence of fungi in the most decomposed manure samples, while fewest species and lowest prevalence were observed in the samples of least decomposed manure. The organic amendment provides the required food sources to the nematode-trapping fungi hence their populations build up. Timm *et al.* (2001) suggested that the increase in nematode-trapping fungi after addition of organic amendment is due to available carbon and energy from the amendment and nitrogen from consumed nematodes. Later Jaffee (2006) reported that alfalfa (*Medicago sativa*) leaves enhanced the populations of *Dactylellina candidum* though other fungal populations were not worked out. A good number of subterranean fungi are natural enemies of plant parasitic nematodes (Nordbring-Hertz *et al.*, 2002). An increase in number of these fungi directly affects the population of nematodes. The devices such as adhesive conidia, branches, knobs and mycelia are used by various fungi to parasitize nematodes (Yang *et al*., 2007). Other fungi immobilize or kill nematodes by releasing toxins (Jansson and Persson, 2000; Sanyal, 2000; Masoomeh *et al*., 2004).

D. Effects on Plants

The yields of potato tubers increased by 27% in the plots treated with either compost or manure while barley yields also increased by 12% when compost was applied (Kimpinski *et al*., 2003). *Sesbania rostrata* and *Aeschynomene afraspera* when used as rotational and green manure crops, increased rice yields due to nutritional effect and the reduction of the populations of *Hirschmanniella mucronata* and *H. oryzae* (Prot *et al*., 1992). Plants treated with poultry and farmyard manures gave significantly higher yields than those of other organic manures. This was because root system was much healthier without galls in poultry and farmyard manure applications and the plants could translocate more nutrients to vegetative organs than heavily-galled roots (Otiefa and Elgindi, 1962). Akhtar and Malik (2000) suggested that organic substrates support high numbers of nematode-destroying fungi and reduced populations of plant parasitic nematodes. Therefore, low numbers of plant parasitic nematodes coupled with high numbers of nematode-destroying fungi and high nutrient levels in the soil have a cumulative effect to plant growth (Wachira *et al*., 2009). Miller and Donahue (1990) reported that organic residues with C: N ratios of 20:1 or narrower have sufficient nitrogen to supply to the decomposing microorganisms and also to release for plant use. Sawdust when applied as a soil amendment, enhanced root-galls in African yam bean in sandy loam soil contrary to the significant reductions in root-galls caused by municipal garbage, swine, compost, poultry and farmyard manures. Poultry and farmyard manures most effectively controlled the disease and improved African yam bean growth and yield (Agu, 2008).

E. Effects on Nematodes

The significant enrichment of soil by manure leads to a 75% increase in total nematode population density at mid-cycle and to a 30% (Villenave *et al*., 2003) increase at harvest time. These changes in nematode community after the application of semi liquid manure are short-lived (Dmowska & Kozlowska, 1988). When incorporated into the soil, organic substrates undergo biologically-mediated decomposition to release NH_4^+, formaldehyde, phenols and volatile fatty acids, among other compounds (Wang *et al.*, 2004). Dmowska and Krassimira (2002) found Rhabditida to be the most dominant group among the bacterivores that increased with application of pig manure and/ or antibiotics. The populations of nematodes particularly bacterivores was significantly reduced (more than 50 %) by earthworm activity both in fresh cow manure and sewage sludge (Domínguez *et al.*, 2003). The effect of earthworms on nematodes may be due to their direct ingestion and digestion (Dash *et al.*, 1980), and consequent effects on the fertility, viability and germination of cysts present in casts (Ellenby 1945; Roessner 1981). The effect may also be indirect, through modifications in soil structure, water regimes, and nutrient cycling processes (Yeates, 1981). Earthworms may decrease total free-living soil nematode populations by up to 66% (Yeates 1981), and plant parasitic nematodes by up to 64 % (Dash *et al.* 1980).

The involvement of soil micro-organisms in nematode control in amended soils has been confirmed by the fact that soil irradiation disrupts the nematicidal effect of these amendments (Kaskavalci, 2007). However, populations of free-living nematodes such as bacterivores, fungivores and predators have been shown to rapidly increase following the addition of organic amendments (Akhtar and Malik, 2000; Jaffee, 2002). It has been, therefore, established that application of organic substrates leads to build-up of micro-organisms which serve as food substrates for free-living nematodes hence their build-up. Poultry manure causes short term (within a growing season) increase in microbiological activity and nutrient cycling, whereas farm yard manure seems to bring long-term changes (Griffiths *et al.,* 1994).

In addition, free-living nematodes may accelerate the decomposition of soil organic matter and increase mineralization of nitrogen and phosphorous thus triggering a chain reaction that favours their build-up (Widmer and Abawi, 2000; Kimenju *et al.*, 2004). Yücel *et al.* (2002) categorised that organic amendments that have high nitrogen content and release ammonia upon decomposition, are more effective in nematode suppression. Unlike the nematode-destroying fungi and the nematode communities which were enhanced by the chicken manure, the growth of the plants was more enhanced by the combination of chicken and cow manures. The increase in growth is attributed to the release of macro- and micronutrients, plant growth regulators and stimulation of beneficial micro flora such as the mycorrhizae fungi (Kaskavalci, 2007). The plants grown on organic amendments were taller and heavier compared to the control (Wachira *et al.*, 2009). The increase in crop vigour may partly be attributed to reduce plant-parasitic nematode populations, but nutrients availability cannot be ignored. Likewise, the decrease in populations of parasitic nematodes has been associated with increased crop yield (Zolda, 2006). Linford *et al*. (1938) were the first to study the nematicidal effects of organic amendments, incorporating chopped pineapple *(Ananas comosus)* leaves into soil for control of *Meloidogyne* spp. in cowpea *(Vigna unguiculata)*. As the population of free-living nematodes increased while that of phytoparasite, *Meloidogyne* spp. decreased, it was suggested that the increased organic matter supported microbial and animal species antagonistic to nematodes. The population of root-knot nematode *Meloidogyne javanica* was significantly reduced due to application of cattle manures with no root galls on tomato roots grown in soils containing 10 or 25% (v/v) cattle manure compost, and very few on those grown in 50% grape marc compost. Significant reduction in galling index was also found on tomato plants grown in soils containing lower concentrations of this compost (Oka and Yermiyahu, 2002). Poultry manure with *Pseudomonas fluorescens* was the best combination for the management of *M. incognita* on tomato but improved management of *M. incognita* can also be obtained in goat dung with *P. fluorescens* or poultry manure with *Azotobacter chroococcum* (Siddiqui, 2004). Some other materials *viz*., pressmud, vegetable- and fruit-processing waste and tobacco wastes were the most effective in reducing the incidence of root knot and development of plant-parasitic nematodes on tomato (Akhtar, 1993.). However, amendments to soil with spent tea, wheat straw, paddy husk, paddy straw, sugarcane trash, domestic garbage, dead vegetation and pigeonpea stubble were also found to be beneficial in nematode control. Some other materials include agricultural wastes in the form of green manures (Crow *et al*., 1995) and dried-crop residues in general and industrial by-products, such as oilcakes, sawdust, cellulosic waste, and sugar-cane bagasse in particular. Besides, other biological wastes (chitin, bone meal, horn meal, sewage-sludge, municipal refuse, and livestock wastes) have been proved successful in controlling nematodes for several years (Akhtar and Alam, 1993). The probable increase in numbers of non-plant parasitic nematodes such as bacterivores, fungivores and omnivores might have contributed to the suppressive characters of the amendments against the plant-parasitic nematodes (Agyarko *et al.,* 2005). The decline in plant parasitic nematodes could probably be attributed to the high number of nematode destroying fungi in the rhizosphere zone. The association between nematode-trapping fungi, organic matter, plant growth and nematodes community is complex. That's the reason it would be difficult to conclude with certainty that the increase of nematode-destroying fungi would lead to automatic reduction of plant parasitic nematodes. Green manures, at high rates, were found to reduce root-knot nematode numbers in the soil. The type of green manure used influenced its effectiveness (Crow *et al.*, 1995). The incidence of Tobacco Rattle virus (TRV) in manured soils was remarkably low due to the affected vector *Paratrichodorus teres* that shows reduced host-finding behaviour (Zoon *et al.*, 2002)

In manure applications density of enrichment opportunists was higher in tropical soils with regard to that found under similar conditions in temperate ecosystems. Furthermore, the abundance of the c-p 2 bacterial-feeding nematodes, belonging mainly to the family Cephalobidae, was strongly correlated with soil microbial biomass. The other c-p feeding guilds showed no correlation with nitrogen flux, or soil microbial biomass. Nematode faunal response to the long-term (20-year) application of nitrogen fertilizer and organic manure was monitored in a Hapli-Udic Cambosol of Northeast China, where no fertilizer (C), nitrogen fertilizer (N), organic manure (M) and nitrogen fertilizer plus organic manure (N + M) treatments were compared. The obtained results showed that total nematode abundance was positively correlated to the M and N + M treatments. The number of *Acrobeloides* sp. increased in the M and N + M treatments at maize jointing and booting stages, and those of *Aphelenchoides* were higher in the N + M treatment than in the N treatment at maize ripening stage. Early in the growing season, the numbers of bacterivores of cp-1 (cp, colonizer-persister) and cp-2 guilds were higher in the M and N+M treatments than in the N treatment, while those of bacterivores and fungivores of cp-4 guilds were higher in the C treatment than in the N, M and N + M treatments. Except at maize seedling stage, the numbers of fungivores of cp-2 and herbivores of cp-3 guilds were higher in the N + M treatment than in the C treatment. Ratios of the weighted abundance of representatives of specific functional guilds were used as indicators of food web structure, enrichment and decomposition channels. Values of the enrichment index (EI) at maize seedling and jointing stages were higher in the M and N + M treatments than in the C and N treatments, while an opposite trend was observed in the channel index (CI). Higher EI and lower CI suggested an enriched soil food web dominated with bacterial decomposition channels in the M and N + M treatments (Liang *et al.*, 2009). The raw (labile form) and composted (non-labile form) manures differ in their chemical composition and effect on microbial activity (Nahar *et al.*, 2006). The volatile compounds of these amendments, especially hydrogen sulfide and methane, have also been shown to be toxic to nematodes (Rodriguez-Kabana *et al.*, 1965; Fortuner and Jacq, 1976). Also they have different effects on soil properties and directly and indirectly affect nematode community structure. Others workers have also shown that low-molecular weight fatty acids, the products of organic manures and amendments are toxic to nematodes (Johnson, 1959; Banage and Visser, 1965; Hollis and Rodriguez-Kabana, 1966; Elmiligy and Norton, 1973; Husain and Masood, 1973; Pillai and Desai, 1975). Similarly, the lethal role of phenols was noted by Taylor and Murant (1966) and Singh and Sitaramaiah (1973) against *Longidorus elongatus* and *Meloidogyne javanica.*

The composting processes overall usually contribute to lowering the availability of metals in amended soil compared to other waste bio-stabilization techniques. The compost matrix modulates metal availability and those low in metals have stronger sorption capacity compared to high metal composts (Smith, 2009). Zinc is the element in sewage sludge-treated agricultural soil identified as the main concern in relation to potential impact. The total metal concentration in compost is important in controlling crop uptake of labile elements, like Zn and Cu, which increases with increasing total content of these elements in compost. The number of opportunistic bacterial-feeding nematodes increases following addition of liquid hog manure and acidified mixtures of the manure to soil. The increase in opportunistic nematodes following manure treatment is due to the increase in their food resources associated with the enrichment of the soil environment with readily degradable compounds. Volatile fatty acids present in the manure persist in the soil for only four days before biological degradation (Mahran *et al.,* 2009). The liquid hog manure is also effective in killing plant-parasitic nematodes.

Poor survival and limited movement of entomopathogenic nematodes in poultry manure appear to make them unlikely candidates for biocontrol of filth flies (Georgis *et al*., 1987). In another study (Bednarek and Gaugler, 1997), organic manure resulted in increased densities of a native population of *Steinernema feltiae*, whereas NPK fertilizer suppressed nematode densities regardless of manure applications.

Oil radish when used as a green manure has dramatically reduced populations of stubby root nematode (*Trichodorus*) and lesion nematode (*Pratylenchus*) in fields (Anon, 2001). Rapeseed and sudan grass green manures grown prior to potatoes provided 72-86% control of the root-knot nematodes (Stark, 1995).

VII. Impact on Environment

Manure nutrient characteristics can vary 30-40 percent from farm to farm depending on the ration fed to animals and the manure handling system. On one hand manure might contribute positively to the environment as it substitutes artificial fertilizer or is used for energy production thereby substituting fossil fuel. Storage and field application systems do have an impact on nutrients available for plant growth.

1. Nutrient Leaching

Nitrogen from manure may be subject to many transformation processes. Fresh animal excreta does not contain N in the form of NO_3. If manure is not aerated by composting, NO_3 will not be formed until after application to the soil. Nitrification or transformation of NH_4 into NO_3, is an aerobic microbiological process that needs oxygen. In well-aerated soil, nitrification of NH_4 added to manure may be completed in a few days. In less aerated soil it may take weeks. At low temperatures, the process is very slow. NO_3 can also be formed after decomposition of the organic N compounds of manure. Transformation of organic N into mineral N is a microbial process called N mineralization. Following NO_3 formation, complete or partial denitrification can take place during which N_2 and N_2O are emitted. Soils may have aerobic and anaerobic conditions occurring simultaneously or alternately as a result of rain showers or irrigation runs. Thus nitrification and denitrification processes may take place both at the same time and alternately. This makes it extremely difficult to predict the fate of applied nitrogen leached as NO_3 or emitted as N_2/N_2O.

Significant nitrogen loss can occur through various storing and handling methods. Temperature influences mineralization of the organic P in manure. It can therefore be assumed that like N, P is mineralized more rapidly in tropical than in temperate conditions. Because the ratio organic P/inorganic P in manure is much lower than organic N/inorganic N, temperature influences the working coefficient of P for single applications much less than that of N. Phosphorus and potassium losses are usually negligible due to their stable nature. The exception would be losses caused by lot/storage runoff and soil erosion (Krawczyk *et al.*, 2005). The harmful effects include transport of nutrients to ground water and surface water bodies through leaching and overland flow. There can also be the accumulation of manure salts under conditions of poor drainage, leading to development of saline and sodic soils. The risk of manure nutrient accumulation and loss can be minimized and the maximum agronomic benefit realized from these nutrients through the use of sound management practices.

Higher soil organic matter content makes soils less susceptible to erosion. Any addition of organic material, including manure contributes to soil organic matter content. The humification coefficient indicates which fraction of the added organic material remains after a certain period of decomposition as stable thus contributing to soil organic matter. Manure has a relatively high humification coefficient, so the contribution of animal manure to the build up of soil organic matter is, therefore, high.

2. Dangerous Volatiles

Due to the considerable biogenic potential of manures their improper use in crops may degrade the soil environment. Gaseous compounds and odours formed adversely affect the natural environment during the storage of manure and when the manure is applied as fertilizer to field grown crops (Schmidt and Bicudo, 2000). GHGs (Green House Gases) are released by cattle manure, in part because of incomplete degradation. For example, digestion in the absence of oxygen creates methane. If feed efficiency is increased, less methane will be produced. All manures have the potential to release GHGs over time whether those are stockpiled or used. Application of more nitrogen through manure may result in excess nitrogen accumulation in soil and increased release of nitrogen as nitrous oxide. Application of manure at the wrong time of year or application of raw manure during wet conditions also increases the release of nitrous oxide. The latter along with ammonia, carbon dioxide, nitrous oxide, nitrate and phosphate, and sulphur hydrogen is responsible for the formation of acid rain and the greenhouse effect. The high concentration of these gases in air also has a negative effect on the health of humans and animals (Kirchmann and Lundvall, 1998), as confirmed by McGinn and Janzen (1998). The effect of organic materials with a high C: N ratio as bedding material, on NH_3 emission is also worth investigating as it may cause microbial immobilization of N from excreta. Often the rate of NH_3 volatilization is influenced by NH_3 concentration in the manure, temperature and wind speed. Volatilization of NH_3, however, is also governed by the chemical equilibrium between NH_4^+ and NH_3 and the equilibrium between aqueous and gaseous NH_3. During storage of manure in the open, NH_3 volatilization and NO_3, P and K leaching occurs particularly when rainfall is high thus resulting in nutrient losses. Likewise, surface spreading of manure during periods of precipitation may lead to NH_3 volatilization, NO_3 and P leaching. Eventually, the agricultural value of animal manure appears to be low. Therefore, uncontrolled use of poultry manure in soil has serious consequences for the natural environment, as it involves over fertilization of soil, nutrient leaching, the resulting contamination of ground waters, and eutrophication of surface waters (Aubert, 2001).

3. Pathogenic Organisms

The poultry manure is found to harbour pathogenic bacteria viz., Enterobacter cloacae, Escherichia coli, Klebsiella ornithiolytica, K. pneumonae, K. terrigina, Proteus vulgaris, Pseudomonas spp., Ps. putrifaciens, Ps. fluorescense, Salmonella spp., S. arizonae and A. hydrophila (Koeypudsa *et al.*, 2005). Diploscapter, a thermotolerant, free-living soil bacterial-feeding nematode commonly found in compost, sewage, and agricultural soil has been reported to transport Salmonella enterica, Escherichia coli and Listeria monocytogenes thus contaminating pre harvest fruits and vegetables (Gibbs *et al.*, 2005). A lack of appropriate hygienic practices in growing, harvesting, processing, distributing, and preparing fruits and vegetables eaten raw can have a substantial impact in increasing the risk of human microbial infections (Beuchat and Ryu, 1997).

4. Antibiotics

Chicken are generally fed with chlortetracycline (CTC) which is excreted in fair concentrations in feaces (Koeypudsa *et al.*, 2005) hence the resulting manure may have serious impacts on the soil biota besides plants. Tainted manure can impact more than just the soil. Once applied to the land, antibiotics can infiltrate water supplies and can seep through the soil into aquifers or spill into surface water due to runoff. Composting reduces concentrations of three antibiotics by 54 percent to 99 percent by degradation, although one drug, sulfamethazine, did not degrade at all.

5. Heavy Metals

Soil contamination with heavy metals and toxic elements due to parent materials or point sources often occurs in a limited area and is easy to identify. Repeated use of metal-enriched chemicals, fertilizers, and organic amendments such as sewage sludge as well as wastewater may cause heavy metal contamination at a large scale. Many chemical processes are involved in the transformation of trace elements in soils, but precipitation-dissolution, adsorption-desorption, and formation of more complex compounds are the most important processes controlling bioavailability and mobility of trace elements in soils. Both deficiency and toxicity of trace elements occur in agroecosystems. Application of trace elements in fertilizers is effective in correcting micronutrient deficiencies for crop production. Remediation of soils contaminated with metals is still costly and difficult although phytoremediation appears promising as a cost-effective approach. Composting processes inherently reduce metal availability compared to other organic waste stabilization methods. Therefore, risks to the environment, human health, crop quality and yield, and soil fertility, from heavy metals in green waste-compost are minimal. Furthermore, composts produced from mechanically-segregated MSW generally contain fewer metals than sewage sludge that is used as an agricultural soil improver under controlled conditions (Smith, 2009). Thus conservative, but pragmatic limits on heavy metals in compost may be set to encourage recycling of composted residuals and contaminant reduction measures, which at the same time, also protect the soil and environment from potentially negative impacts caused by long-term accumulation of heavy metals in soil.

VIII. Management of Manure

Animal manure can be an effective source of nutrients for crop growth and should be viewed as a resource rather than a waste product. Like commercial inorganic fertilizers, they also can provide economic returns without adversely impacting the environment when managed properly. The main problems encountered with manure application are the transport of nutrients to ground water and surface water bodies through leaching and runoffs; escape of manure-derived gases such as ammonia and nitrous oxide into the atmosphere and accumulation of manure salts under conditions of poor drainage, leading to development of saline and sodic soils.

A. Field Mapping and Crop Needs

Developing field maps showing current soil fertility and plans for future crops will help identify the fields requiring manure application. Manure needs to be applied to the crops that use the nutrients most efficiently viz., nitrogen is used more efficiently by corn and cereal crops compared to legumes. If manure is applied to meet the crop nitrogen demand, the phosphorus and potassium applied will eventually be in excess in soil. However, the legume hay crop rotation may bring down the levels of P and K, thus providing a closer balance of the three nutrients in due course.

Selection of field should be based on topography of the land, besides, the location to streams and water bodies should also be considered while selecting the field to apply manure. Steeply sloping lands should not be selected for manure application unless the material can be immediately incorporated to avoid any possibility of contamination due to surface water runoff. The grass buffering strip between all fields and streams is another important strategy to prevent the flow of nutrient due to erosion or run off.

B. Reducing Energy Inputs

For a proper and efficient manure management the energy inputs should be minimized. The selection of manure for a field should be based on the soil type and the crop nutritional requirements of N, P and K. The area of contact between manure and air should be minimized by better or more frequent cleaning of the stables and by covering the manure storage. Another strategy can be to decrease the concentration of NH_3 (and thus the vapour pressure) in the manure by dilution with water or to lower the pH in the manure by addition of acid and to stop the formation of NH_3 from uric acid in poultry manure by drying until the dry matter content exceeds 70%. The straw supplement increases the efficiency of composting chicken manure, reduces ammonia and carbon dioxide emissions, and increases the reduction of biogenic potential. Land application of raw or composted manure can be tailored to reduce the emission of GHGs and their impact on the environment.

C. Optimising Manure Application

Timing manure application correctly and ensuring proper application amounts can contribute to an overall reduction in GHG emissions from agricultural operations. The manure can be spread in calculated amount evenly over the entire field. Technology for application of liquid manure has progressed rapidly in the past few years. A substantial reduction in NH_3 volatilization after application is possible by using one of the following techniques:

1. Experiments with 1:1 and 1:3 water diluted manure have shown substantial reductions in emission (Monteny and Erisman, 1998).
2. An artificial shower of 20 mm produced by means of a sprinkler installation.
3. Immediate tillage after manure application in such a way that the manure is worked in.
4. The liquid manure is applied in very narrow furrows made in the sward.
5. The manure is injected into the soil at a depth of 10-20 cm to get good distribution of nutrients for uniform access by plants. Injection of manure reduces odour and has been shown to increase crop use of N compared to its surface placement.
6. Drag-hose machine applies the manure very close to the surface in such a way that contact between manure and air is minimal at the time of application.

Similar to commercial inorganic fertilizer N, application of manure in colder season reduces conversion of the ammonium N to nitrate and helps prevent losses by leaching and denitrification. In the case of solid manure, suitable technology has not yet been developed for placement of the solid manure below the soil surface with minimal disturbance. Broadcasting of solid manure is therefore usually followed by incorporation with a tillage implement to reduce volatilization losses of ammonia. For solid cattle manure with significant amounts of straw, immediate versus delayed (24 hour) incorporation did not have a large influence on nitrogen recovery. However, volatilization losses associated with delayed incorporation will be higher when the manure contains a higher proportion of fecal material and nitrogen in the ammonium form.

Both liquid and solid manures provide nutrients in fairly low concentrations by volume when compared to a commercial fertilizer. As energy costs continue to grow so is the cost of commercial fertilizers used for crop production. Thus manure applications can be considered as a prime source of N, P and K for the growing crops. The nitrogen-deficient sheep dung, horse dung and rice straw amendments should be supplemented with an inorganic nitrogenous fertilizer in order to counteract their high carbon/nitrogen ratio. This could improve the C/N ratio, thus helping to eliminate the undesirable effects of nitrogen starvation to plants while maintaining their benefit for nematode control. As a result, current limitations to more widespread use are mainly in transportation costs from the site of the manure production. Composting and other processing techniques that help reduce the volume of manure will increase the economic hauling distance. Composting has recently garnered significant attention as a method of manure management. High-temperature composting of manure, designed to kill pathogens, is required for crops with organic label. That could also eliminate some antibiotics (Dolliver and Gupta, 2008).

Conclusion

Manure may contribute positively to the environment as a substitute of artificial fertilizer and also a substitute of fossil fuel. Its storage and field application systems do have an impact on nutrients available for plant growth. Climate also plays crucial role as the nutrients N, P are mineralized more rapidly in tropical than in temperate conditions. Animal manure with a relatively high humification coefficient, though builds up soil organic matter with great pace, however, their improper use in crops may degrade the soil environment. The manures release green house gases over time causing a negative effect on the health of humans and animals. Also, lack of appropriate hygienic practices in handling the produce obtained from manure treated fields, can have a substantial impact in increasing the risk of human microbial infections. The antibiotics as well as unwanted nutrients can infiltrate water supplies or contaminate the ground water table. Nevertheless, when managed properly manures can provide economic returns without causing much adverse impact to the environment. The risk of manure nutrient accumulation and loss can be minimized and the maximum agronomic benefits realized from these nutrients through the use of sound manure nutrient management practices. Such plans should include location of farm, maps, manure inventory and provisional statements relating to handling of manure/ livestock waste.

REFERENCES

Agyarko, K. & Asante, J.S. (2005). Nematode dynamics in soil amended with neem leaves and Poultry manure. *Asian Journal of Plant Sciences* 4: 426-428.

Agyarko, K., Kwakye, P.K., Bonsu, M., Osei B.A & Asante, J.S. (2005) Effect of Neem and organic amendments on nematode populations in a coastal Savanna tropical soil *Phytoparasitica* 33: 343-346.

Agu, C. M. (2008). Effects of organic manure types on root-gall nematode disease and African Yam bean yield. *Medwell online Agricultural Journal* 3: 14-16.

Akhtar, M. (1993). Utilisation of plant-origin waste materials for the control of plant-parasitic nematodes. *Bioresource Technology* 46: 255-257

Akhtar, M. & Alam, M..M. (1993) Utilization of waste materials in nematode control: A review. Bioresource Technology 45: 1-7

Akhtar, M. & Malik, A. (2000). Roles of organic soil amendments and soil organisms in the biological control of plant-parasitic nematodes: a review. *Bioresource Technology* 74: 35-47.

Anderson, J. M. 1995. The soil system. In: Global Biodiversity Assessment: Section 6. Mooney, H.A., Lubchenco, J. Dirzo, R. & Sala, O.E. (Eds.) pp 406–412. Cambridge Univ. Press, Cambridge, UK.

Anon. (2001). Oil radish green manure continues promise against nematodes. The Grower. June–July. 7.

Axelsen, J. A. & Kristensen, K. T. (2000). Collembola and mites in plots fertilised with different types of green manure. *Pedobiologia* 44: 556-566.

Banage, W. B. & Visser, S. A. (1965). The effect of some fatty acids and pH on a soil nematode. *Nematologica,* 11: 255-262.

Bednarek, A. & Gaugler, R. (1997) Compatibility of soil amendments with entomopathogenic Nematodes. *Journal of Nematology* 29: 220–227.

Beuchat, L. R. & Ryu, J. H. (1997). Produce handling and processing practices. *Emerging Infectious Diseases* 3: 459-465.

Binet, F. & Trehen, P. (1992). Experimental microcosm study of the role of Lumbricus terrestris (Oligochaeta: Lumbricidae) on nitrogen dynamics in cultivated soils. *Soil Biology and Biochemistry* 24: 1501–1506.

Brodie, B.B. & Good, J.M. (1973). Relative efficacy of selected volatile and non volatile nematicides for control of Meloidogyne incognita on tobacco. *Journal of Nematology* 5: 14-18.

REFERENCES

Agyarko, K. & Asante, J.S. (2005). Nematode dynamics in soil amended with neem leaves and Poultry manure. *Asian Journal of Plant Sciences* 4: 426-428.

Agyarko, K., Kwakye, P.K., Bonsu, M., Osei B.A & Asante, J.S. (2005) Effect of Neem and organic amendments on nematode populations in a coastal Savanna tropical soil *Phytoparasitica* 33: 343-346.

Agu, C. M. (2008). Effects of organic manure types on root-gall nematode disease and African Yam bean yield. *Medwell online Agricultural Journal* 3: 14-16.

Akhtar, M. (1993). Utilisation of plant-origin waste materials for the control of plant-parasitic nematodes. *Bioresource Technology* 46: 255-257

Akhtar, M. & Alam, M..M. (1993) Utilization of waste materials in nematode control: A review. Bioresource Technology 45: 1-7

Akhtar, M. & Malik, A. (2000). Roles of organic soil amendments and soil organisms in the biological control of plant-parasitic nematodes: a review. *Bioresource Technology* 74: 35-47.

Anderson, J. M. 1995. The soil system. In: Global Biodiversity Assessment: Section 6. Mooney, H.A., Lubchenco, J. Dirzo, R. & Sala, O.E. (Eds.) pp 406–412. Cambridge Univ. Press, Cambridge, UK.

Anon. (2001). Oil radish green manure continues promise against nematodes. The Grower. June–July. 7.

Axelsen, J. A. & Kristensen, K. T. (2000). Collembola and mites in plots fertilised with different types of green manure. *Pedobiologia* 44: 556-566.

Banage, W. B. & Visser, S. A. (1965). The effect of some fatty acids and pH on a soil nematode. *Nematologica,* 11: 255-262.

Bednarek, A. & Gaugler, R. (1997) Compatibility of soil amendments with entomopathogenic Nematodes. *Journal of Nematology* 29: 220–227.

Beuchat, L. R. & Ryu, J. H. (1997). Produce handling and processing practices. *Emerging Infectious Diseases* 3: 459-465.

Binet, F. & Trehen, P. (1992). Experimental microcosm study of the role of Lumbricus terrestris (Oligochaeta: Lumbricidae) on nitrogen dynamics in cultivated soils. *Soil Biology and Biochemistry* 24: 1501–1506.

Brodie, B.B. & Good, J.M. (1973). Relative efficacy of selected volatile and non volatile nematicides for control of Meloidogyne incognita on tobacco. *Journal of Nematology* 5: 14-18.

Brodie, B.B., Good, J.M. & Jaworski, C.A. (1970). Population dynamics of plant nematodes in cultivated soil: Effect of summer cover crops on newly cleared land. *Journal of Nematology* 2: 217-222.

Chitwood D.J. (2002). Phytochemical based strategies for nematode control. Annual Review of *Phytopathology* 40: 221-249.

Coderre, D., Estevez, B., Cote, D., Page, F.(1995).Short-term effects on earthworm populations of sidedressing corn with solid dairy cattle manure in a boreal climate. *American Journal of Alternative Agriculture* 10: 50-73.

Crow, W. T., Guertal, E. A. & Rodriguez, R. (1995). Green manures: an environmentally friendly way to control root-knot nematodes. *Highlights of Agricultural Res*earch *42.*

Curry, J.P. (1976). Some effects of animal manures on earthworms in grassland. *Pedobiologia,* 16: 425-438.

Dash, M. C., Senapati, B. K. & Mishra, C. C. (1980) Nematode feeding by tropical earthworms. *Oikos* 34, 322–325.

Daulton, R.A.C. (1963). The behaviour and control of root knot nematode, *Meloidogyne javanica* in tobacco as influenced by crop rotation and soil fumigation. *Proceedings of Third World Tobacco Science Congress*. pp10.

Devliegher, W. & Verstraete, W. (1997). The effects of *Lumbricus terrestris* on soil in relation to plant growth: Effects of nutrient enrichment processes (NEP) and gut-associated processes (GAP). *Soil Biology and Biochemistry* 29: 341-346.

Dmowska, E. & Kozlowska, J. (1988). Communities of nematodes in soil treated with semi-liquid manure. *Pedobiologia* 32: 323-330.

Dmowska, E. & Krassimira I. (2002) Reaction of bacterial-feeding nematodes to pig manure with antibiotic and manure without antibiotic. Fourth International Congress of Nematology Programme and Abstracts 8–13 June. Tenbel, La Galletas, Arona, Tenerife, Canary Islands, Spain.

Dolliver, H. & Gupta, S. (2008). Antibiotic losses in leaching and surface runoff from manure-amended agricultural land. *Journal of Environmental Quality* 37: 1227-1237.

Domek-Chruscicka, K. & Seniczak, S. (2005). The effect of pig liquid manure fertilization on the crop of alternating grassland and some groups of soil mesofauna. *Folia Biologica (Cracow) Supplements* 53: 139-143.

Dominguez, J., Edwards, C.A. & Subler, S. (1997) A Comparison of Vermi-composting and Composting. Biocycle, April.

Dropkin, V. H. (1980). Introduction to Plant Nematology. John Wiley & Sons, New York, NY. 256p.

Kirchmann H. & Lundvall, A. (1998). Treatment of solid animal manures. Identification of low NH3 emission practices. *Nutrient Cycling in Agroecosystem* 51: 65-71.

Koeypudsa, W., Yakupitiyage, A. & Tangtrongpiros, J. (2005). The fate of chlortetracycline residues in a simulated chicken-fish integrated farming systems. *Aquaculture Research* 36: 570-577.

Krawczyk, W., Herbut, E., Walczak, J., Muchacka, R. & Radecki, P. (2005). The environmental impact of composting manure ISAH vol 2- Warsaw, Poland.

Liang, W., Lou, Y., Li, Q., Zhong, S., Zhang, X. & Wang, J. (2009). Nematode faunal response to long-term application of nitrogen fertilizer and organic manure in Northeast China. *Soil Biology and Biochemistry* 41: 883-890.

Linford, M. B., Yap, F. & Oliveira, J. M. (1938). Reduction of soil populations of root-knot nematode during decomposition of organic matter. *Soil Science* 45:127-141.

Mackay A.D., Kladivko E.J. (1985). Earthworms and the rate of breakdown of soybean and maize residues in soil. *Soil Biology and Biochemistry* 17: 851-857.

Mahran, A., Tenuta, M., Lumactud, R.A. & Daayf, F. (2009). Response of a soil nematode community to liquid hog manure and its acidification. *Applied Soil Ecology* 43: 75-82

Mahoney, C. J. & Strongman, D. B. (1994). Nematophagous fungi from cattle manure in four states of decomposition at three sites in Nova Scotia, Canada. *Mycologia*, 86: 371-375.

Masoomeh, S.G., Mehdi, R.A., Shahrokh, R.B., Ali, E. & Rasoul, Z. (2004). Screening of soil and sheep faecal samples for predacious fungi: Isolation and characterization of the nematode-trapping fungus *Arthrobotrys oligospora*. *Iranian Biomedical Journal* 8: 135-142.

McGinn, S.M. & Janzen, H.H. (1998). Ammonia sources in agriculture and their measurement. *Canadian Journal of Soil Science* 78: 139-148.

Miller, R. W. & Donahue, R. L. (1990). Organic matter and container media. Soils: An introduction to Soils and Plant Growth. 6th (Ed). Prentice Hall, Inc., Englewood Cliffs, N. J., U.S.A. 181-225.

Minor, M. A. & Norton, R. A. (2004). Effects of soil amendments on assemblages of soil mites (Acari: Oribatida, Mesostigmata) in short-rotation willow plantings in central New York. *Canadian Journal of Forest Research* 34: 1417–1425. doi:10.1139/x04-035

Minton, N. A. & Donnelly, E. D. (1971). Reaction of field grown lespedeza to selected *Meloidogyne* spp. *Journal of Nematology* 3: 369-373

Monteny, G.J. & Erisman, J.W. (1998). Ammonia emission from dairy cow buildings: a review of measurement techniques, influencing factors and possibilities for reduction. *Netherlands Journal of Agricultural Science* 46: 225-247.

Nahar, M.S., Grewal, P.S., Miller, S.A., Stinner, D., Stinner, B.R., Kleinhenz, M.D., Wszelaki, A. & Doohan, D. (2006). Differential effects of raw and composted manure on nematode community, and its indicative value for soil microbial, physical and chemical properties. Applied Soil Ecology 34: 140-151.

Nordbring-Hertz, B., Jansson, H.B. & Tunlid, A. (2002). Nematophagous Fungi: Encyclopedia of Life Sciences. Macmillan Publishers Ltd., London.

Oka, Y., Nacar, S., Putieusky, E., Ravid, U., Zohara, Y. & Spiegal, Y. (2000). Nematicidal activity of essential oils and their components against the root knot nematode. *Phytopathology* 90: 710-715.

Oka, Y. & Yermiyahu, U. (2002). Suppressive effects of composts against the root-knot nematode *Meloidogyne javanica* on tomato. *Nematology* 4: 891-898.

Otiefa, B. A. & Elgindi, O. N. (1962). Influence of parasitic duration of Meloidogyne javanica on host nutrient uptake. *Nematologica* 8: 216-220.

Parmelee R.W., Beare M.H., Cheng W., Hendrix P.F., Rider S.J., Crossley D.A., Jr., Coleman D.C. (1990). Earthworms and enchytraeids in conventional and no-tillage agroecosystems: A biocide approach to assess their role in organic matter breakdown. *Biological Fertility of Soils* 10:1-10.

Parmelee R.W. & Crossley D.A., Jr. (1988). Earthworm production and role in the nitrogen cycle of a no-tillage agroecosystem on the Georgia piedmont. *Pedobiologia*;32: 353-361.

Pillai, *S*.N. & Desai, M.V. (1975). Anthelminthic property of "marotti" cake *(Hydnocarpus laurifolia). Pesticides* 9: 37-39.

Prot, J., Soriano, I. R. S., Matias, D. M. & Savary, S. (1992). Use of Green Manure Crops in Control of *Hirschmanniella mucronata* and *H. oryzae* in Irrigated Rice. *Journal of Nematolgy* 24: 127-132.

Quarles, W. (1993). Rapeseed green manure controls nematodes. The IPM Practitioner. April. p. 15.

Rodriguez-Kabana, R., Jordan, J. W. & Hollis, J . P. (1965). Nematodes : Biological cont,rol in rice fields : Role of hydrogen sulfide. *Science,* 148: 524-526.

Roessner, J. (1981) Einfluss von Regenwürmern auf phytoparasitäre Nematoden. *Nematologica* 27: 339-347.

Sanyal, P.K. (2000). Screening for Indian isolates of predacious fungi for use in biological control against nematode parasites of ruminants. *Vet. Res. Commun.* 24: 55-62.

Sasser, J. N. (1990). Plant-parasitic Nematodes: The Farmer's Hidden Enemy. North Carolina State University Press, Raleigh, NC. 47-48.

Schmidt, D. R. & Bicudo J. R. (2000). Odor and hydrogen sulfide emissions during the composting of caged layer manure. *Mat of Sec. Int. Conf., Des Moines*, Iowa, 75-83.

Siddiqi, M.R., 2000. Tylenchida Parasites of Plants and Insects. 2nd Edn., CAB International, Wallingford, UK., ISBN: 085199-202-1, 848p.

Siddiqui, Z. A. (2004) Effects of plant growth promoting bacteria and composed organic fertilizers on the reproduction of *Meloidogyne incognita* and tomato growth. *Bioresource Technology* 95: 223-227.

Singh, R. S. & Sitaramaiah, K. (1973). Control of plant parasitic nematodes with organic amendments of soil. *Experimental Station & College of Agriculture, Pantnagar, India, Research Bulletin* 6: 289.

Smith, S. R. (2009). A critical review of the bioavailability and impacts of heavy metals in municipal solid waste composts compared to sewage sludge. *Environment International* 35: 142-156.

Stark, J.C. (1995). Development of sustainable potato production systems for the Pacific Northwest. SARE Final Report.

Stirling, G.R. (1991). Biological Control of Plant Parasitic Nematodes. CAB International, Wallingford, UK. 275 p.

Taylor, C. E. & Murant, A. F. (1966). Nematicidal activity of aqueous extracts from raspberry canes and roots. *Nematologica,* 12 : 448-449.

Timm, L., Pearson, D. & Jaffee, B. (2001). Nematode trapping fungi in conventionally and organically managed corn-tomato rotations. *Mycologia* 93: 25-29.

Villenave, C., Bongers, T., Ekschmitt, K., Fernandes, P. & Oliver, R., (2003). Changes in nematode communities after manuring in millet fields in Senegal. *Nematology* 5: 351-358.

Wachira, P.M., Kimenju, J.W., Okoth, S.A. & Mibey, R.K. (2009) Stimulation of Nematode-Destroying Fungi by Organic Amendments Applied in

Management of Plant Parasitic Nematodes. *Asian Journal of Plant Sciences* 8: 153-159. DOI: 10.3923/ajps. 153.159.

Walker, G. E. (2004). Effects of *Meloidogyne javanica* and organic amendments, inorganic fertilisers and nematicides on carrot growth and nematode abundance. *Nematologia Mediterranea,* Vol. 32, No. 2, pp. 181-188.

Wang, K. H., McSorley, R. & Gallaher, R. N. (2004). Effect of *Crotalaria juncea* amendment on squash infected with *Meloidogyne incognita. Journal of Nematology* 36: 290-296.

Wang, K. H., Sipes, B. S. & Schmitt, D. P. (2002). Suppression of *Rotylenchulus reniformis* by *Crotalaria juncea, Brassica napus*, and *Target erecta. Nematropica.* 31: 237-251.

Widmer, T.L. & Abawi, G.S. (2000). Mechanism of suppression of *Meloidogyne hapla* and its damage by a green manure of Sudan grass. *Plant Disease* 84: 562-568.

Williams, G. & Williams, P. (1990a). Sesame residues *vs*. harmful nematodes. HortIdeas. March. p. 35.

Williams, G. & Williams, P. (1990b). (Some) plant nutrients repel harmful nematodes. HortIdeas. June. p.63.

Williams, G. & Williams, P. (1993). Wheat vs. nematodes causing peach tree short life. HortIdeas. July. p.76.

Yang, Y., Yang, E., Zhiqiang, A.& Liu, X. (2007). Evolution of nematode-trapping cells of predatory fungi of the Orbiliaceae based on evidence from rRNA-encoding DNA and multiprotein sequences. *Proceedings of National Academy of Science* 104: 8379-8384.

Yasuhiro, ITO. (1971). Predation by manure-inhabiting mesostigmatids (Acarina: Mesostigmata) on some free-living nematodes. *Applied Entomology and Zoology* 6: 51-56.

Yeates, G.W. (1981) Soil nematode populations depressed in the presence of earthworms. *Pedobiologia* 22: 191–195.

Yücel, I., Shuttleworth, W.J., Pinker, R. T., Lu, L. & Sorooshian, S. (2002). Impact of Ingesting Satellite-Derived Cloud Cover into the Regional Atmospheric Modeling System. *Monthly Weather Review* 130: 610-628.

Zolda, P. (2006). Nematode communities of grazed and ungrazed semi-natural steppe grasslands in Eastern Austria. *Pedobiologia* 50: 11-22.

Zoon, F.C., van Bruggen, A.H.C., de Heij, A., Asjes, C.J. & van den Ende, J.E. (2002). organic amendments on incidence of nematode-borne tobacco rattle virus. In: VIII International Symposium on Flower bulbs, Cape Town, South Africa. *Acta Horticulturae* 570: 287 - 292.

In: Manure: Management, Uses and… ISBN: 978-1-61668-424-2
Editors: C. S. Dellaguardia, pp. 87-118

Chapter 3

PHOSPHORUS FORMS IN ANIMAL MANURE AND THE IMPACT ON SOIL P STATUS

Zhongqi He[1*] and Zhengxia Dou[2]
[1]USDA-ARS, New England Plant, Soil, and Water Laboratory, Orono, ME 04469
[2]University of Pennsylvania, Center for Animal Health and Productivity, Kennett Square, PA 19348

ABSTRACT

Increased knowledge of manure phosphorus (P) chemistry is needed to optimize nutrient recycling and to minimize adverse environmental effects associated with land application of manure. Part I of this chapter reviews and discusses various manure P characterization methods including sequential fractionation, solution ^{31}P nuclear magnetic resonance (NMR) spectroscopy, enzymatic hydrolysis, and solid state spectroscopic techniques. We then present a case study that compares P forms in conventional and organic dairy manure identified by solution and solid state ^{31}P NMR spectroscopy. The results of this case study suggest that the lability of P may differ between organic and conventional dairy manures; therefore specific management practices may be required

* Corresponding author: Email address: Zhongqi.He@ars.usda.gov, USDA-ARS, New England Plant, Soil, and Water Laboratory, University of Maine, Orono, ME 04469. Tel: (207) 581-3373. Fax: (207) 866-0464. Trade names mentioned in the paper are for information only and do not constitute endorsement, recommendation, or exclusion by the USDA-ARS.

for organic dairy manure. Part II of the chapter focuses on the impacts of manure application on soil P status. For this part we present and discuss the results of laboratory aerobic incubation and field studies to evaluate short- and long-term impacts, respectively. The lack of stable organic P in relevant extracts of the laboratory or field soils with manure amendment indicates possible degradation of stable manure P. Sequential fractionation of pasture soils reveals that the levels of labile P are more closely related to the number of years receiving poultry litter, whereas the levels of stable P are related to the cumulative amount of poultry litter applied. The two different patterns imply that the impact of manure application on the environment (including runoff potential) should be gradually factored in with increasing length of application.

INTRODUCTION

A significant amount of phosphorus (P) is present in animal manure. Land application of animal manure can benefit crop production by providing P and other nutrients. However, P runoff from agricultural soil due to over-application of animal manure may cause surface water degradation. Best management strategies are needed to optimize recycling manure P while minimizing adverse consequences of manure land application. The chemical composition of P in animal manure is one of the critical factors controlling P transport and bioavailability in soils. In the last decade, efforts have been made to increase basic knowledge of manure P chemistry by applying novel and/or modified advanced analytic methodologies. In this chapter we review and discuss the research on characterization of P forms in animal manure. We further demonstrate, using case studies, how this knowledge has improved our understanding of the fate and transformations of animal manure in agricultural soils.

I. CHARACTERIZATION OF PHOSPHORUS IN ANIMAL MANURE

I.1. Sequential Fractionation of Manure P

Sequential fractionation studies of manure P initially focused on P availability for plants. Using sheep manure, McAuliffe and Peech (1949) were the first to prescribe a fractionation procedure which involved extracting the sample with an alcohol/ether (3:1) mixture, followed by 5% trichloroacetic acid (TCA) extraction, and then ashing the residues. The corresponding P fractions were designated as lipid P, acid soluble inorganic and organic P, and residual P. This procedure was later used by Peperzak et al. (1959) and Barnett (1994a; 1994b) to analyze P in various manures such as dairy cows, feeder and finisher cattle, hogs, and broiler and layer poultry. Of total P in the manure samples, lipid P accounted for 0.4 to 2.1%, acid soluble inorganic P 35 to 63% and organic P 8 to 53%, with residual P ranging from 11 to 41%. The acid extracted P is considered available for plant uptake (Barnett, 1994a; 1994b).

More recently, a more sophisticated sequential fractionation procedure was adapted and utilized featuring increasingly stronger chemical solutions - deionized water, 0.5 M $NaHCO_3$, 0.1 M NaOH, and 1.0 M HCl, in that order - to quantify pools of P with varying degrees of solubility. This procedure was based on the Hedley (1982a) procedure for soil P fractionation but replaced the anion exchange resin extraction as the first step for soil P fractionation with deionized water for animal manure because the latter contains a considerable fraction of water soluble P (Dou et al., 2000; He and Honeycutt, 2001). Classes of P in manure are defined by the extractant's removal from the manure matrix, i.e. H_2O-P, $NaHCO_3$-P, NaOH-P, HCl-P, as well as residual P (He et al., 2003a). The five P fractions are generally assigned to labile water soluble P, labile P adsorbed on crystalline surfaces, labile P associated with exteriors of Al and Fe oxides or carbonates, stable Ca-bound P, and recalcitrant P, respectively, but such assignments should not be considered as exclusive (He et al., 2003a). These functional P categories can be further divided into inorganic P (P_i) and organic P (P_o). The latter is generally calculated as the difference between P_i and total P (P_t) in an extractant. It has been argued (Haygarth and Sharpley, 2000) that some labile polyphosphate and P_o could be hydrolyzed during the assay and thus become part of P_i determined by the Mo blue method (Murphy and Riley, 1962). Thus, some researchers prefer the term "molybdate-reactive P" or "soluble reactive P" (SRP) for the P_i determined by this colorimetric method (Haygarth and Sharpley, 2000). Indeed, this term change only superfically improves the method. If the sponatanous hydrolysis is severe, a modified Mo blue method could be used to address the concern (Dick and Tabatabai, 1977; He and Honeycutt, 2005).

Although time-consuming, the sequential extraction procedure allows P to be categorically quantified, thereby providing a tool for investigating P transformations in animal manure affected by various environmental and/or management factors. Phosphorus forms in various animal manures have been investigated using this procedure, including from cattle (Ajiboye et al., 2004; Dou et al., 2002; He et al., 2004b; Sharpley and Moyer, 2000), swine (Ajiboye et al., 2004; He et al., 2004a; Wienhold and Miller, 2004); poultry manure/litter (Dou et al., 2002; He et al., 2004b; He et al., 2006c; Sharpley and Moyer, 2000); and other less common types of manure such as from deer, sheep, fox, and mink (McDowell and Stewart, 2005; Uusitalo et al., 2007).

The modified Hedley procedure for characterizing manure P follows the same principle generalized in Figure 1 but with variations in certain aspects. The ratio of manure and extractant (g:mL) differed depending on the investigators: 1:200 (Sharpley and Moyer, 2000), 1:100 (Dou et al, 2000; Wienhold and Miller, 2004; He et al, 2006c), 1:50 (He et al, 2004b; Uusitalo et al., 2007). Extraction time is 16 h in most studies. However, using a dairy and a poultry manure sample, Dou et al. (2000) found that P release into the water extract fraction was not greatly affected by shaking times from 1 to 16 h. Of the sum of water extractable P removed through repeated (multiple) 1-h extractions, 68% (dairy manure) and 75% (poultry manure) was released in the first 1-h extraction, which was designated as "readily soluble P" and believed to be most susceptible to potential environmental loss (Dou et al., 2000). As 1-h water-extractable P in manure has been proposed as an indicator of runoff potential of manure P (Kleinman et al., 2002), a 1-h extraction time in the sequential fractionation scheme could make the data from water extractable P in the sequential fractionation more compatible with those for runoff potential, particularly if the water:manure ratio and the condition of the manure (i.e, dried or fresh) are further unified. Vadas et al. (2006) have found that the water:manure ratio and whether or not manures are dried and ground or fresh make big differences in how much P can be extracted by water. The 16-h extraction time in other steps not only improves the extraction efficiency (Dou et al., 2000), it is also convenient operationally as a researcher can set up the extraction overnight. Unlike soil samples, animal manure extraction residues do not form tightly-compacted pellets during centrifugation. Hence, it is difficult to remove all clear supernatant (extracts) from the centrifuge tube. He et al. (2006c) suggested that a subsequent washing of the residues with 5 mL of water could reduce the "contamination" of the supernatant leftover in the next extraction.

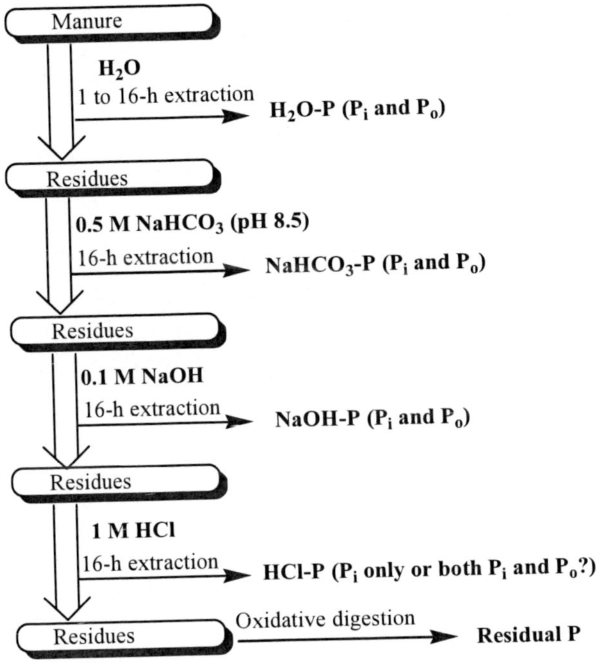

Figure 1. A typical modified Hedley sequential fractionation procedure for animal manure P extraction. Please see the main text for discussion

There is concern regarding the forms of P (inorganic vs. organic) in the HCl extract. It is generally accepted that for soils, HCl extracts contain P_i only whereas P_o is negligible (Hedley et al., 1982b; Cross and Schlesinger, 1995). However, He et al. (2006f) demonstrated with 15 soil and 8 manure samples that some manure HCl extracts did contain significant amounts of organic P although other manure samples did not. Two manure samples even had P_o greater than P_i. Their data indicate that P_o in HCl fractions should not be ignored but determined experimentally. Without considering HCl P_o, any data using the sequential fractionation schemes might be incomplete.

Figure 2 displays P distribution patterns in four animal manures. If P in the H_2O and $NaHCO_3$ fractions is considered labile, more than 70% of total P in the dairy and swine manures was labile. In contrast, only about 30% of total P was labile in the poultry litter and fox manure. The labile P was primarily inorganic, whereas 54, 43 and 90% of P in the NaOH extracts were P_o in dairy manure, swine manure, and poultry litter, respectively. Organic P also accounted for 40 and 60% of P in the HCl fractions of swine manure and poultry litter. Phosphorus in fox manure had a pattern similar to poultry litter.

Unlike poultry litter, there was little or no P_o in the NaOH and HCl fractions of fox manure. Foxes consume bone meal which has a high level of sparingly soluble P_i (Ylivainio et al., 2008). Even though P contents in animal manure vary greatly due to different management factors, such as feedstuffs, bedding, and storage type, similar distribution patterns have been observed for the same type of manure by different research groups. For example, the total P content of swine manure was 3.7 g kg^{-1} of dry matter in He et al. (2003a) and 19.8-33.6 kg^{-1} of dry matter in Wienhold and Miller (2004). The P in the swine manure in He et al (He et al., 2003a) was distributed as 47.2%, 19.1%, 18.3%, 12.2% and 3.1% of total P in the five fractions, very similar to the pattern of swine manure in Figure 2. Uusitalo et al. (2007) and Dail et al. (2007) reported similar distribution patterns for dairy and poultry manures, respectively. Therefore, best P management strategies should consider the specific manure P solubility (stability) in each type of manure.

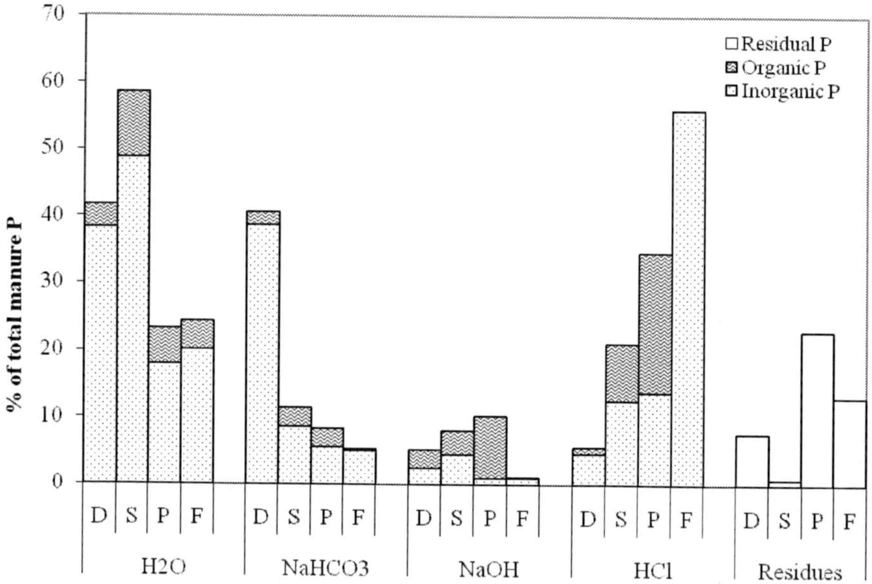

Figure 2. Phosphorus distribution patterns in sequentially extracted H_2O, 0.5 M $NaHCO_3$ (pH 8.5), 0.1 M NaOH, 1 M HCl and residual fractions of animal manure. Data are adapted from He et al. (2004b) for dairy manure (D) with 9.2 g P kg^{-1} of dry matter; Wienhold and Miller (2004) for swine manure (S) with 25.8 g P kg^{-1} of dry matter; He et al. (2006c) for poultry litter (P) with 22.3 g P kg^{-1} of dry matter; and Uusitalo et al. (2007) for fox manure (F) with 34.3 g P kg^{-1} of dry matter.

The manure P fractionation based on a modified Hedley sequential procedure has been applied to study how dietary manipulation in animal feeding may affect P characteristics and their environmental implications. Using fecal samples from three independent dairy feeding trials and sequential fractionation, Dou et al. (2002) demonstrated that decreasing dietary P levels not only led to less excretion of total P in feces, but more importantly it reduced the amount and proportion of P that was water soluble. For instance, when dietary P was reduced from 6.7 to 5.1 or 3.4 g P kg^{-1} feed DM, water soluble P in feces decreased from 10.46 to 7.13 or 2.91 g kg^{-1} fecal DM, respectively. Water soluble P accounted for 83, 77, or 56% of fecal total P produced from the three diets. Other P fractions remained relatively small and were unaffected by dietary P concentration. These findings have significant environmental implications. Water soluble P is most susceptible to potential losses through surface runoff and subsurface drainage. Reducing excess P in diets would markedly decrease potential environmental impacts. The P sequential fractionation procedure has also been used to investigate the efficacy of various amendments added to manure to stabilize P. For example, in a laboratory study, Dou et al. (2003) found that compared to unamended controls, aluminum sulfate ($Al_2(SO_4)_3$, alum) effectively reduced water soluble P in amended dairy, swine, or poultry litter manures. The reduction ranged from 80 to 99% at treatment rates of 100 to 250 g alum kg^{-1}manure DM. Fluidized bed coal combustion fly ash reduced water soluble P by 50 to 60% at a rate of 400 g kg^{-1} for all three manures, whereas coal flue gas desulfurization by-product reduced water soluble P by nearly 80% when added to swine manure and broiler litter at 150 and 250 g kg^{-1} but was ineffective in dairy manure. Another by-product, anthracite coal refuse fly ash, was ineffective for all three manures. Sequential extraction results indicated that the by-product treatments shifted manure P from water extractable P into a less vulnerable fraction, $NaHCO_3$-P, while the alum treatment shifted the P into even more stable forms, mostly NaOH-P. Such shifts in P fractions would have little influence on P availability for crops over the long term but may retard and reduce potential losses of P immediately following manure applications.

I.2. Phosphorus Forms Identified by Solution ^{31}p NMR Analysis

As ^{31}P is the naturally occurring phosphate isotope, solution ^{31}P NMR spectroscopy is an excellent tool for the study of P forms in animal manure. Leinweber et al. (1997) reported the solution ^{31}P NMR spectra of 0.1 M and

0.5 M NaOH extracts of swine and chick manures. More recently, NaOH-EDTA, adapted from a soil organic P extraction procedure (Bowman and Moir, 1993), has been used to extract P in animal manure for solution ^{31}P NMR analysis (Cade-Menun, 2005b). This method allows P to be grouped into six types of P forms: phosphonates, orthophosphate, orthophosphate monoesters, orthophosphate diesters, pyrophosphate, and polyphosphate (Table 1). Within these compound classes, several specific P compounds can be further identified, such as phytate (*myo*-inositol hexakisphosphate, IP6) in the orthophosphate monoester range, and DNA in the orthophosphate diester region. Although chemical shift values of many other P model compounds have been reported in the literature (Cade-Menun, 2005a; Makarov, 2005), caution is needed when attempting to apply those data to interpretation of the ^{31}P NMR spectra of "real world" manure samples, because ^{31}P NMR chemical shift values depend on the ionic strength of the solution, pH, and the presence of paramagnetic (such as Fe, Cu, and Mn) ions. In practice, identifying every P compound may be of limited use because the P classification presented in Table 1 should provide sufficient information with respect to the bioavailability or plant nutritional value of manure P.

Table 1. Major phosphorus forms in NaOH-EDTA extracts of cattle manure (CM), poultry litter/manure (PM), and swine manure (SM) identified by solution ^{31}P NMR spectroscopy

Form	Chemical shift (ppm)	Relative abundance (%)		
		CM[a]	PM[b]	SM[c]
Phosphonate	19 to 24	~ 1	ND[d]	NR[e]
Orthophosphate	5.7 to 6.1	62-70	34-66	86-94
Phosphomonoester:	3 to 6:	15-28:	30-66:	5-12:
Phytate		5-6	24-51	trace
Others		14-20	7-12	
Phosphodiester:	-1.0 to 2.5:	4-11:	2-7:	0-2:
DNA		1-5	0-1	
Lipids		2-6	2-6	
Others		< 1	<1	
Pyrophosphate	-4.5 to -5.0	1-7	0-1	0-2
Polyphosphate	-19 to -21	1-8	ND	<1

[a]: Data are from (He et al., 2007c; Turner, 2004); [b]: Data are from (He et al., 2007c; Hill and Cade-Menun, 2009; Turner, 2004); [c]: Data are from (Leytem et al., 2004; Turner, 2004); [d]: ND, not detected or less than 0.01% of extractable P; [e]: NR, not reported.

Although solution ^{31}P NMR spectroscopy identifies the same P forms in different types of animal manure, the abundance of these forms is different among manures (Table 1). In NaOH-EDTA extracts of cattle manure, the majority of P is in the forms orthophosphate (62-70%), orthophosphate monoesters (15-28%), and phosphodiesters (4-11%). Pyrophosphate, polyphosphate, and phosphonates are also present with a total amount less than 10%. The NaOH-EDTA extracts of poultry manure are dominated by both orthophosphate (34-66%) and orthophosphate monoesters (30-66%) (He et al., 2007c; Hill and Cade-Menun, 2009; Turner, 2004). Phytate (24-51%) is the major orthophosphate monoester form (He et al., 2007c; Hill and Cade-Menun, 2009). Interestingly, recent reports (Leytem and Thacker, 2008; Leytem et al., 2004; Turner, 2004) showed that swine manure contains mainly P_i with trace phytate P irrelevant to the phytate content in feeds even though pigs are monogastric animals that are known for not being able to digest phytate P. Further research with various diets is needed to confirm the observation (Leytem and Thacker, 2008).

It should be kept in mind that NaOH-EDTA solution may be of low efficacy in extracting P from some manure samples, e.g. those with high Ca contents (He et al., 2007c). For example, NaOH (0.25 M)-EDTA (0.05 M) extracted only 48-70% of P in cattle, deer and sheep dung collected from an Australian grazing pasture (McDowell and Stewart, 2005). Turner (2004) reported a recovery of 80% of P from a cattle manure using the NaOH (0.25 M)-EDTA (0.05 M). With less than 100% recovery, caution must be taken when interpreting P forms in NaOH-EDTA extracts as whole manure P composition. He et al. (2007c) used an additional 1.0 M HCl extraction to recover the residual P after NaOH (0.25 M)-EDTA (0.05 M) extraction. Of the total P in their poultry manure samples, 65-71% was released in NaOH-EDTA solution while 1.0 M HCl recovered an additional 33-34%. Solution ^{31}P NMR spectroscopy revealed that, of the HCl-extracted P, 22-23% was phytate, 4-5% was other monoester P and 0-1% was pyrophosphate P. This work indicates that the residual part of NaOH-EDTA extraction should also be characterized for a full understanding of manure P composition if NaOH-EDTA extraction does not recover all or nearly all P from a manure sample. Solution ^{31}P NMR spectroscopy coupled with sequential fractionation could be an alternative to provide comprehensive information on manure P composition and solubility (Ajiboye et al., 2007; He et al., 2008b; Turner and Leytem, 2004)

Table 2. Distribution of P in swine and cattle manures revealed by enzymatic treatment. Data are (adapted from (He and Honeycutt (2001))

P_o form	Swine manure fraction				Cattle manure fraction			
	H_2O	$NaHCO_3$	NaOH	HCl	H_2O	$NaHCO_3$	NaOH	HCl
	% [a]	%	%	%	%	%	%	%
Inorganic P	61.1	38.9	18.5	88.3	57.8	52.5	35.0	92.4
Simple monoester	2.8	2.0	35.0	1.7	0.3	0.5	9.8	1.3
Phytate-like	15.2	2.0	4.2	2.5	7.2	0	0	0
DNA-like	2.1	0.7	1.5	- [b]	5.1	4.2	2.1	-
Pyrophosphate	1.6	0	0.2	-	6.8	0	0	-
Nonhydrolysable	18.6	56.6	40.8	7.7	23.2	42.3	54.6	6.1

[a]: Percentage of P in each fraction; [b]: Not determined.

I.3. Characterization of Organic P by Enzymatic Hydrolysis

Enzymes are proteins that catalyze reactions to cleave specific chemical bonds. In nature, phosphatases catalyze the hydrolysis of certain organic P compounds to produce inorganic orthophosphate. Because organic P must be hydrolyzed to inorganic P prior to uptake by plants and microorganisms, enzymatic hydrolysis provides an estimate of the hydrolyzable, and thus the bioavailable, organic P in environmental samples, which is not possible to obtain with NMR analysis. He et al. (2009c) recently reviewed the principles and applications of enzymatic hydrolysis to investigate various environmental samples; therefore this section focuses only on the lability and forms of P in animal manure revealed by enzymatic hydrolysis.

He and Honeycutt (2001) first examined the release of P_o in sequentially extracted fractions of swine manure and cattle manure. About 49% of P in swine manure and 44% in cattle manure were present in organic forms, mainly distributed in the H_2O, $NaHCO_3$, and NaOH fractions. After pH adjustment, a portion of these fractions was incubated separately with alkaline phosphatase and wheat phytase, or incubated first with nuclease P1 or pyrophosphatase followed by wheat phytase. The enzymatic hydrolysis treatment did provide information about P forms found in the sequential fractions, in addition to the conventionally determined values of P_i and P_o (Table 2). The four designated types of P_o were distributed unequally in the sequential extracts. Phytate-like P was most prevalent in the H_2O fraction and accounted for 39% of P_o in swine manure and 17% of P_o in cattle manure. This finding is consistent with the lack of phytase enzymes in swine digestive systems (Wodzinski and Ullah, 1996). Simple orthophosphate monoesters comprised the major P_o form identified in the NaOH fractions. A small percentage of DNA-like P was found in each fraction (1.1-12%). Organic pyrophosphate was found only in the H_2O extracts.

As various approaches to phosphatase hydrolysis have been used to evaluate the lability of organic P in environmental samples, the diversity of approaches has complicated data comparison (He et al., 2009c). A unified approach for enzyme hydrolysis would allow data comparison across a range of P sources and forms. He et al. (2004a;2004b) have developed a stepwise approach for this purpose. In this enzymatic approach, organic P extracted from manure or soil is incubated at pH 5.0 (100 mM sodium acetate buffer) with (i) acid phosphatase from potato; (ii) acid phosphatase from potato and nuclease P1; or (iii) phosphatases from potato and wheat germ plus nuclease P1. Each of the enzymes shows optimal activity at 37°C and near pH 5. The P differentially released by these three enzymes is designated as simple monoester P, polynucleotide-like P, and phytate-like P, respectively. Another advantage of the modified method is that possible errors due to incomplete substrate overlaps by separate incubations with an individual enzyme are minimized by the stepwise addition of enzymes.

This enzymatic approach, coupled with sequential fractionation, provides a novel tool for understanding manure P chemistry and its impacts on soil P levels (He et al., 2004c; He et al., 2006c; Waldrip-Dail et al., 2009). A significant finding for application of this coupled method is that HCl fractions in the Hedley fractionation may contain both P_i and P_o , rather than P_i only as assumed in some of the literature (He et al., 2006f). In the original procedure proposed by Hedley et al. (1982a; 1982b), no P_o was included in the HCl fractions because the amount of P_o in the HCl fractions of their samples was negligible based on preliminary measurement. Since then, many researchers have simply followed the procedure and assumed that no P_o was present in HCl fractions without any measurement (Cross and Schlesinger, 1995). In practice, the inconsistent and/or different background interferences between P_i and P_t measurements made estimation of concentrations of P_o less accurate, even sometimes negative, as P_i is generally measured by a molybdenum blue colorimetric method, and P_t is measured using the same method after oxidative digestion, or by an inductively-coupled plasma method with or without digestion (Ajiboye et al., 2007; Codling, 2006; He et al., 2006f). Addition of enzyme(s) basically keeps the medium matrices of extracts unchanged. Thus, increases in P_i determination after enzyme incubation should be attributed to the hydrolysis of P_o by the enzymes added to these extracts. Using enzymatic hydrolysis, He et al. (2006f) measured P_o in 15 soil and eight manure samples.

Their data indicated that, whereas the HCl fractions of some samples contained negligible amounts of P_o, others contained significant amounts of P_o. The concentrations of P_o were greater than P_i in one soil sample and two manure samples. Apparently, it should be determined experimentally whether the HCl fraction of a sample contains P_i only or both P_i and P_o. Without considering the HCl P_o species, any data using the Hedley sequential fractionation schemes might be incomplete or less accurate. Inclusion of HCl-extractable P_o with other sequentially extracted P forms would provide more accurate and comprehensive information on the fate and interchanges of P forms in soil and animal manure under various environmental conditions (He et al., 2006f).

Since both ^{31}P NMR spectroscopy and phosphatase hydrolysis are valuable tools for P characterization, He et al. (2007c) quantitatively compared the P forms in NaOH (0.25 M)-EDTA (0.05M) extracts of fresh and dried dairy and poultry manure determined by the two methods (Figure 3). Enzymatic hydrolysis identified polynucleotide-like phosphodiesters with abundances of 6.4, 3.6, 3.4 and 4.3% of total extracted P in the four manure samples, respectively. The abundance quantified by solution ^{31}P NMR spectroscopy was 7.7, 7.2, 4.2, and 3.4% for total orthophosphate diester P, and 1.3, 0.8, 0.9, and 1.6% of total extracted P in the four manure samples for DNA, respectively. These data are reasonable since the abundance of polynucleotide-like P should be greater than that of DNA but less than that of total orthophosphate diester P. The other three types of P identified by the two methods were more similar (Figure 3). Although the P forms and quantities detected by the two methods were not always the same, comparison of the two sets of data indicated that the patterns of P distribution in these animal manure samples determined by the two methods were similar, with a correlation coefficient of 0.937 (Figure 3). He et al. (2008b) further examined the differences in solution ^{31}P spectra of sequential fractions of poultry litter before and after enzymatic hydrolysis. The peaks of P_o species became very weak or disappeared in the solution ^{31}P NMR spectra of the enzymatically-treated fractions, which confirmed the validity of P identities in the untreated fractions detected by enzymatic hydrolysis or solution ^{31}P NMR spectroscopy. These observations suggest that both methods can be used to characterize P in animal manure and it is acceptable to compare the information on P data obtained by the two methods although some discretion should be applied. Researchers can utilize the method that best fits their specific research goals, or use both methods to obtain a full spectrum of manure P characterization.

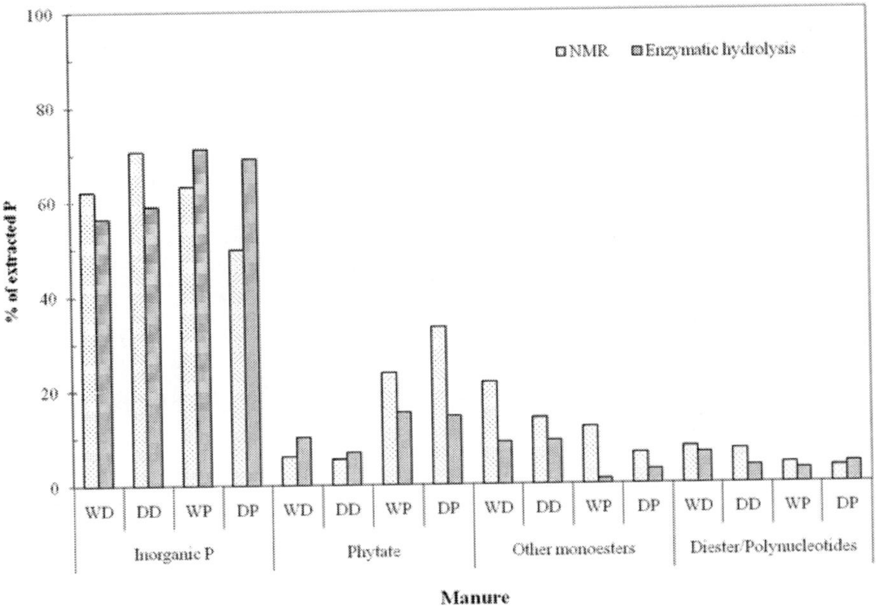

Figure 3. Comparison of phosphorus forms in NaOH (0.25 M)-EDTA (0.05 M) extracts of animal manure identified by enzymatic hydrolysis and solution ^{31}P NMR analysis (He et al., 2007c). WD-wet dairy manure, DD-dry dairy manure, WP-wet poultry manure, DP, dry poultry manure. The correlation coefficient of the two sets of data is 0.937 with the statistic significance at $\alpha < 0.001$

The NaOH extractant used in either the sequential fractionation or solution ^{31}P NMR spectroscopy could spontaneously hydrolyze some labile P_o compounds (He et al., 2003b; Pant et al., 1994). To obtain more accurate information on labile P_o, He et al. (2003b; 2006d) separated a manure sample into soluble and residual parts under mild assay conditions using water and sodium acetate buffer (pH 5.0). The labile P_o in the two parts was then released by enzymatic hydrolysis. Using the mild assay method, He et al. (2003b) investigated the effects of storage on inorganic P (P_i), enzymatically hydrolysable organic P (P_e), and non-hydrolysable organic P (P_{ue}) in cattle manure and swine manure. Their data indicated that the major change during the storage of animal manure for a year was the increase in P_{ue}; thus, manure P solubility may increase with storage, but the increase would not produce more bioavailable P in the manure. The mild assay conditions could be further modified to investigate specific P compounds. For example, the reducing agent sodium dithionite has been included in the incubation mixtures to explore reducible Fe(III)-associated P_i and P_o in poultry litter and dairy manure (He et al., 2006a). Dao et al. (2006) directly added EDTA and fungal phytase to diluted dairy manure to characterize mildly (mainly Ca and Mg) complexed P_i and P_o. The simplified method can also be used as a reliable and convenient analytic method to directly quantify the phytate concentration in animal manure (Dao and Hoang, 2008; He et al., 2009a), since the use in feeds of low-phytate grains and supplementation with phytase enzymes may result in more unpredictable phytate concentration in animal manure (Leytem and Thacker, 2008; Maguire et al., 2004; Wienhold and Miller, 2004).

I.4. Metal Speciation of P Identified by Solid State Spectroscopic Techniques

The forms and lability of P in animal manure have been investigated by solution-based techniques as discussed above, in which the metal speciation of P was estimated by the properties of the solution and/or the extractant. Recently, solid-state based advanced spectroscopic techniques, mainly solid state ^{31}P NMR and X-ray absorption near edge structure (XANES) spectroscopy, have been used to characterize the metal species of P in environmental samples (Ajiboye et al., 2007; Gungor et al., 2007; Hunger et al., 2005; Jayasundera et al., 2005). Those efforts have mainly focused on metal species of inorganic phosphate. Hunger et al. (2004) characterized the solid state ^{31}P NMR spectral features of poultry litter amended with alum, finding a mixture of poorly ordered wavellite and phosphate surface complexes on aluminum hydroxide that had been formed by alum hydrolysis. Furthermore, phosphate associated with Al comprised 40% of P with 14% of P as calcium phosphate phases in the alum-amended poultry litter. Shand et al. (2005) reported the main peaks and spinning sidebands of dicalcium phosphate dihydrate and ammonium magnesium phosphate hexahydrate in the solid state ^{31}P NMR spectra of sheep feces. Jayasundera et al. (Jayasundera et al., 2005) proposed the presence of various P_i and P_o compounds (including P_i complexed with Ca) in dairy manure samples they investigated by solid state ^{31}P NMR spectroscopy.

There are more published manure P research studies using XANES spectroscopy than solid state ^{31}P NMR spectroscopy. Shober et al. (2006) applied XANES fitting processes to identify various P species present in biosolids and manures, such as hydroxylapatite, phosphate sorbed to Al hydroxides and on ferrihydrite, β-tricalcium phosphate and phytate.

Phosphorus K-edge XANES spectra of poultry manure showed no evidence of crystalline P minerals but instead showed the dominance of soluble Ca-P species and free and weakly bound phosphates (Sato et al., 2005). However, diets incorporating modified P minerals could change P speciation in manure: less soluble hydroxylapatite was observed and the aqueous phosphate contribution disappeared in P K-edge XANES spectra of turkey manure with Ca to P ratios greater than 1.7 produced from modified diets (Toor et al., 2005). Using X-ray powder diffraction and XANES, Gungor et al. (2007) determined P speciation of raw and anaerobically digested dairy manure with an emphasis on the Ca and Mg phosphate phases. Linear combination fitting of the XANES spectra showed that 57 and 43% of P in the raw manure was associated with anhydrous dicalcium phosphate and struvite ($MgNH_4PO_4 \cdot 6H_2O$), respectively. In the anaerobically digested manure, 78% of P was present as struvite and 22% of P was associated with hydroxylapatite, suggesting P speciation shifts toward Mg and the least soluble Ca orthophosphates following anaerobic digestion.

Whereas these two solid state spectroscopy methods have provided novel insight into metal P_i speciation, in studies where researchers have used one or two phytate compounds (i. e. sodium phytate $Na_{12}IP6$ and/or Ca hydrogen phytate $CaH_{10}IP6$) as reference compounds to study phytate, these two techniques were not able to distinguish phytate compounds in environmental samples (Beauchemin et al., 2003; Peak et al., 2002). He et al. (2006e; 2007a; 2007b) prepared different metal phytate compounds, and studied their solid state ^{31}P NMR and P K-edge XANES spectral features. These spectra demonstrated significant differences in the shapes and positions of spectral peaks, indicating these two spectroscopic techniques could be used to advance our knowledge of metal speciation of phytate in animal manure. In addition, Kruse et al. (2009) recently reported the P $L_{2,3}$-edge XANES spectra of a wide range of reference P compounds. Their results demonstrated that these P $L_{2,3}$-edge spectral features are richer than those of the P K-edge spectra. Therefore, the application of P $L_{2,3}$-edge XANES spectroscopy may yield more reliable results on P speciation in animal manure.

I.5. Case Study: P Forms in Conventional and Organic Dairy Manure

Organic dairy production has increased rapidly in recent years. Organic dairy farms have significant differences from conventional counterparts, including fewer imports of protein and energy feeds, a higher proportion of forage crops in the ration, and an increased reliance on manure and compost as nutrient sources, rather than chemical fertilizers. Although these differences could impact the turnover and availability of manure P to plants, little information is available to aid organic dairy farmers in making nutrient and manure management decisions. For this reason, He et al. (2009d) comparatively characterized P in organic (OD) and conventional (CD) dairy manure by multiple extractions, solution and solid state ^{31}P NMR spectroscopic techniques. Phosphorus in both types of dairy manure was separately extracted with water, Na acetate buffer (100 mM, pH 5.0) plus 20 mg Na dithionite mL^{-1} (NaAc-SD), or 0.025 M NaOH with 50 mM EDTA (NaOH-EDTA). The first extractant is for the most labile P (Kleinman et al., 2007; 2002), the second has a greater affinity for Fe-associated P (He et al., 2006a), and the last one extracts stable metal-P species (Cade-Menun, 2005b; He et al., 2007c). Whereas the three extractions gave data on solubility (i. e. lability) of manure P, further characterization of the extracts and residues by solution and solid state ^{31}P NMR spectroscopy provided information about the P species in the manures.

Whereas water extracted more P from CD (50.3%) than OD (42.4%), NaAc-SD and NaOH-EDTA were more efficient in extracting P from OD than CD (Table 3). These data imply that there is both readily labile and recalcitrant P present in CD, but more moderately labile P in OD. Solution ^{31}P NMR spectra show seven types of P species present in both types of dairy manure (Table 3). Orthophosphate (i. e. P_i) accounts for 70 to 89% of extracted P in these extracts. Polyphosphate and pyrophosphate are not present in water extracts, but in both NaAc-SD extracts, and in the NaOH-EDTA extracts of OD. As NaAc-SD extracted significantly more Fe than did NaOH-EDTA; this suggested that polyphosphate and pyrophosphate are associated with Fe in these samples (He et al., 2009d). Recently, Dou et al. (unpublished data) also observed greater percentages of pyrophosphate and Fe in NaAc-SD than in NaOH-EDTA extracts of two dairy manure samples. These observations demonstrate that the mild NaAc-SD solution is an alternative to NaOH-EDTA for extracting P from samples where P is more closely associated with Fe and hydrolysis of condensed P compounds is a concern.

In the solid state ^{31}P NMR spectra of the unextracted manures, a sharp peak appeared at 6.2 ppm with a broad peak centered at 2.6 ppm with minor spinning sidebands (He et al., 2009d). In contrast, the spectra of the residues of water and NaAc-SD extractions showed a major asymmetric complex peak at a region from -0.2 to 1.7 ppm with stronger spinning sidebands. The strength of spinning sidebands is a critical factor distinguishing organic and inorganic compounds with similar major isotropic chemical shifts: strong spinning side bands are observed only in the spectra of metal phytate compounds, minor spinning sidebands in the spectra of hydrogen inorganic phosphate compounds and no detectable spinning sidebands in the spectra of tribasic phosphate compounds (He et al., 2007a; 2007b). With these spinning sideband features and deconvolution of the major isotropic chemical shift peaks, various Na/K, Ca, Mg, Zn and Al species of P_i and P_o compounds have been identified from these solid state ^{31}P NMR spectra. Combining the data from the multiple extractions and solution ^{31}P NMR spectroscopy, 15 P species have been identified in these two manure samples (Table 4). Thus, He et al. (2009d) concluded that their conventional dairy manure sample contained relatively higher contents of soluble P_i species and recalcitrant metal P species. In contrast, their organic dairy manure sample contained more moderate soluble (labile) Ca- and Mg-P species. As these moderate forms would gradually release the nutrients, organic dairy manure could act as a slow-release fertilizer, thus increasing the efficiency of organic dairy manure as a nutrient source while reducing the potential of runoff into nearby surface or groundwater sources. However, further testing is required, since only one organic dairy manure sample has been studied to date.

Table 3. P in extracts of conventional (CD) and organic (OD) dairy manure by H_2O, 100 mM sodium acetate buffer (pH 5.0) with 20mg/ml sodium dithionite (NaAc-SD) and 0.05M NaOH-0.025M EDTA (NaOH-EDTA) by solution ^{31}P NMR spectroscopy. Data are (adapted from (He et al. (2009d))

Sample	P_t	P_i	Phytate	Other Monoesters	DNA	Other Diesters	PyroP	PolyP
	% of total manure P							
CD- H_2O	50.3	41.1	0.0	5.9	0.8	2.5	0.0	0.0
CD-NaAc	88.2	61.9	6.4	9.3	1.4	1.4	5.6	2.1
CD-NaOH	81.9	55.9	7.5	11.1	1.2	2.5	1.2	2.5
OD-H_2O	42.4	37.5	0.0	2.8	0.3	1.7	0.0	0.0
OD-NaAc	94.9	73.2	10.0	6.6	0.9	1.7	1.7	0.9
OD-NaOH	96.9	77.2	9.5	6.3	0.8	1.6	1.6	0.0

II. Impacts of Manure Application on Soil P Status

II.1. Case Study: Laboratory Aerobic Incubation to Evaluate the Short-Term Impacts

Laboratory aerobic incubation has been used to evaluate the mineralization and bioavailability of N and P in soils amended with animal manure (Abbasi et al., 2007; Eghball et al., 2005; Griffin et al., 2005; Preusch et al., 2002). Sequential fractionation can be used to monitor changes in soil P during the incubation experiments (Buehler et al., 2002; Qian and Schoenau, 2000) and more information on labile P_o can be obtained when enzymatic hydrolysis is coupled with the sequential fractionation (He et al., 2004a). With this coupled approach, He's group conducted laboratory aerobic incubation experiments to evaluate the short-term impacts of dairy and poultry manure addition on soil P status (He et al., 2004c; 2006b; Waldrip-Dail et al., 2009). They measured P_i, P_{ho}, and P_{nho} (enzymatically hydrolyzable and nonhydrolyzable P_o, respectively) in the sequentially-extracted fractions of manure-amended soil at different incubation times and analyzed their patterns of change over time. Since the mechanism of interconversion of P species can be complicated, He et al. (2004c) assumed that interconversion of P species among the different fractions is mainly a physicochemical process and transformation of different P species (P_i, P_{ho}, and P_{nho}) within in a given fraction are mainly mediated by biological and biochemical processes. On the basis of this assumption, He et al. (He et al., 2004c; 2006b; Waldrip-Dail et al., 2009) assessed the P dynamics of two Maine soils in response to dairy and poultry manure application.

An unnamed sandy loam and a Lamoine silt loam were used in the laboratory aerobic incubation experiments, with 11 dairy manure samples plus two chemical fertilizer controls (NH_4Cl and KH_2PO_4) (He et al., 2004c; 2006b). The P species in the H_2O, 0.5 M $NaHCO_3$ (pH 8.5) and 0.1 M NaOH extracts of soils were quantified over the 108-day incubation period. P in 1.0 M HCl was not determined, since these dairy manures contained only a very small amount of HCl-extracted P (He and Honeycutt, 2001; He et al., 2004a); therefore,this fraction should not significantly impact soil P dynamics. Data from this incubation experiment demonstrated complementary fluctuations between P_i in the $NaHCO_3$ and NaOH fractions as well as between labile P (P_i and P_{eo}) and P_{nho} within the same NaOH fraction. He et al. (2004a) proposed that these complementary fluctuations could be an important mechanism for maintaining the balance between labile and immobile P in soils. Furthermore, the general distribution patterns in manure-amended soils for the P_i, P_{ho}, and P_{nho} fractions followed the trend of changes for the inorganic fertilizer-amended (control) soils, indicating that soil properties played a major role in controlling P dynamics. In the H_2O and $NaHCO_3$ fractions, soil bioavailable P (P_i and P_{eo}) was linearly related to the added P. However, NaOH extractable soil P was not closely related to the P applied in the manure. In other words, a single application of dairy manure impacted soil bioavailable P in a manner similar to fertilizer and did not significantly impact soil P properties and the environment. This means growers converting from conventional to organic practices (i.e. where manure is the sole source of N and P) may estimate manure bioavailable P to meet their short-term P needs whereas the impact of manure application on soil properties and environment should be gradually factored in with increasing years of application (He et al., 2006b).

A significant portion of P in poultry manure/litter is HCl-extracted P, which differs from the relative P fraction ratios of dairy manure (Figure 2). Thus, the aerobic incubation of poultry manure-amended soils provides information on the fate of the stable HCl-extracted P in poultry manure upon application to soils (Waldrip-Dail et al., 2009). Overall, poultry manure amendment affects the soil P status primarily in the H_2O- and HCl-P_i fractions. As observed with dairy manure, soil properties, including soil mineral phase and sorption-desorption processes, are key factors in controlling P dynamics in soils amended with poultry manure. With up to 58% of extractable P in the HCl fraction of the poultry manure, which had relatively equal amounts of P_i and P_o (Dail et al., 2007), the P_i level in the HCl fraction of poultry manure-amended soils remained relatively stable over the course of the incubation. In contrast, the P_{nho} fraction rapidly became enzymatically hydrolysable, with a portion remaining in the HCl fractions and some transformed to other fractions. This observation indicates that the stable HCl-extractable P in poultry manure has rapidly degraded to more labile P forms in these amended soils. Whereas sequentially extracted HCl P in soil is considered less plant available (Cross and Schlesinger, 1995), the incubation data from this experiment confirmed that this may not be true for the HCl-extracted P in animal manure (Dail et al., 2007). Therefore, soil amendment with moderate levels of poultry manure does not appear to cause accumulation of stable P forms, and that the large HCl-P fraction (31.1% of total P) in poultry manure could act as a source of plant-available P over a short term (e. g. a growing season) (Waldrip-Dail et al., 2009). A pot experiment with rye grass is currently underway to further test the fate of PL P in soils with plant growth.

Table 4. P species in the conventional (CD) and organic (OD) dairy manure estimated by the combination of selective extractions (1), solution (2) and solid-state (3) ^{31}P NMR spectroscopies. Data are adapted from He et al. (2009d)

P species	Percentage of total P		Identification method
	CD	OD	
Na/K HPO_4^{2-} bonded with H_2O	16.9	24.0	1, 2 and 3
Soluble Ca- and Mg- PO_4^{3-}	24.2	13.5	1, 2 and 3
Fe-related pyrophosphate	5.6	1.7	1, 2 and 3
Moderate soluble hydrogenated Ca-PO_4^{3-}	20.8	-[a]	1, 2 and 3
Moderate soluble tribasic Ca-P minerals	-[a]	35.7	1, 2 and 3
Polyphosphate	2.5	0.9	1 and 2
Soluble DNA	0.8	0.3	1 and 2
Moderate soluble DNA	0.6	0.6	1 and 2
Other diesters (soluble)	2.5	1.7	1 and 2
Moderate Mg and Zn hydrogenated phytate species	6.4	10.0	1, 2 and 3
Other soluble monoesters	5.9	2.8	1 and 2
Moderate soluble other monoesters	3.4	3.8	1 and 2
Al phytate species, some Al phosphate possible	2.0	1.6	1, 2 and 3
Stable divalent metal phytate and phosphate	13.5	6.9	1, 2 and 3
Phosphonates	trace	2.4	3

[a]: Minor moderate hydrogenated Ca-PO_4^{3-} species or tribasic Ca-P minerals might be present.

II.2. Case Study: P Speciation in Heavily Manured Soils

Soils receiving long-term manure applications typically have P built up to levels that far exceed the agronomic optimum required for satisfactory crop production. Such P-enriched soils are often associated with elevated P loss to waters, contributing to accelerated eutrophication of surface waters. Using the Hedley sequential extraction procedure, Sharpley et al. (2004) compared 20 agricultural soils that received poultry, dairy, or swine manure for >10 yr to similar soils receiving no manure or fertilizer. They reported an overall change in P chemistry in the manured soils, with considerable shifts from NaOH extractable to HCl extractable inorganic P in manured soils as compared to the untreated soils. Employing solution ^{31}P NMR spectroscopy, Dou et al. (2009)

recently determined P functional groups in some of the samples used in Sharpley's study (2004). The ^{31}P NMR results (Table 5) showed that orthophosphate was the primary form of P in manured (79-93% of total extract P) and untreated soils (33-71%). Pyrophosphate was identified in all soils, whereas polyphosphate was present in seven out of nine manured soils (9-47 mg P kg^{-1} and <2%) but absent in the untreated soils. Organic P accounted for as little as 5.3% of extracted P in one soil that received duck manure, or as high as 60.9 % in another soil that received no P application. Of organic P forms, orthophosphate diesters were detected in only a few soils in relatively small proportions (<3% of extracted P); these soils included both treated and untreated soils. The majority of organic P was in the form of orthophosphate monoesters, with phytate constituting 17 to 43% of the monoesters. Nevertheless, levels of phytate were similar in manured soils (52–116 mg P kg^{-1}, 2–8%) and untreated soils (43–137 mg P kg^{-1}, 6–22%), suggesting little accumulation of NaOH-EDTA extractable phytate despite long-term manure applications. Poultry manure is known for containing high concentrations of phytate. Estimated amount of phytate applied in the relevant soils would be in the magnitude of 300 kg P ha^{-1} (Dou et al., 2009), translating to about 450 mg phytate P kg^{-1} for the top 5 cm of soils or 150 mg kg^{-1} for the top 15 cm of soils. Clearly, the amounts of phytate detected in these soils (52, 69, 86, 95, and 116 mg P kg^{-1}) were considerably less than the estimated range, suggesting a lack of quantifiable accumulation. The lack of phytate accumulation in manured soils was also reported by He et al (2008b) and Hill and Cade-Menun (2009).

What could have happened to manure-origin phytate in these soils? Part of it may have transferred from the soils in runoff, or it may have translocated downward, and/or leached out of the soil profile. Leaching loss of organic P has been reported (Toor et al., 2003). It is generally believed that phytate molecules have a high affinity for interaction with soil constituents such as (hydr)oxides of Al and Fe, leading to immobilization (Guan et al., 2006; He et al., 2006a; 2006e). However, for P-saturated soils, such immobilization may be limited, rendering manure-borne phytate susceptible to potential leaching or runoff loss. Another possibility is that the phytate was transformed into inorganic orthophosphate or other chemical forms, as suggested in the short laboratory incubation described in section II-1 (Waldrip-Dail et al., 2009). Leytem et al. (2006) also showed that phytate was no longer detectable after a 12-wk incubation of poultry manure-amended calcareous soils while Olsen P in the soils increased. While the observations in both Dou et al. (2009) and Leytem et al. (2006) suggested that degradation of phytate P was occurring in soils, it must be pointed out that profiling of soil P composition based on solution ^{31}P in NaOH-EDTA extracts alone may not always be accurate.

II.3. Case Study: Different Accumulation Patterns of Labile and Stable P Fractions in Pasture Soil after Long-Term Poultry Litter Application

A long-term field experiment in which poultry litter was applied to pasture soils has been maintained for two decades in the Sand Mountain region of north Alabama, USA (He et al., 2008a; 2009b). Five fields received poultry litter (PL) continuously for 0 (control), 5, 10, 15, and 20 years at rates of 2.27, 2.27, 3.63, and 1.36 Mg ha^{-1} yr^{-1}, respectively. Previous work (He et al., 2009b) observed that the P in these five fields had increased with increasing cumulative PL application, suggesting that residual P from PL had accumulated in these soils. Thus, He et al. (unpublished data) further investigated the P accumulation patterns of this residual PL P by sequential fractionation. In the surface soil (0-20 cm), repeated PL application significantly increased P in all five pools of these pasture soils (Figure 4). However, there were apparently two different impact patterns. The levels of labile P (P_i or P_t) extracted by water or $NaHCO_3$ were more closely related to the number of years of poultry litter application. In contrast, the stable NaOH-extractable P in the surface soil was built up, with a closer relationship to the cumulative applied amount or annual rate. No statistically significant correlation was observed between the HCl-extractable and residual P and the three application parameters. These findings suggest that the PL P accumulated in these soils had undergone transformation and interchanges among the five P fractions or pools. The most apparent change was in the HCl-extractable P_o, which accounted for 21% of total P in PL (Figure 2) but was not present in the HCl fractions of litter-applied soils. The four soluble sequential fractions of soils with 0-yr and 20-yr PL applications were also subjected to solution ^{31}P NMR analysis (He et al., 2008b). The only P species detected in the HCl fractions of both soils was orthophosphate. These observations indicated that PL P accumulated in these soils had gone through transformation and interchanges among the five P fractions or pools in these pastures field as observed in the laboratory aerobic incubation.

Table 5. Phosphorus functional groups in NaOH-EDTA extracts of soils (receiving 8-10 yr manure applications vs. untreated) as percentage of total peak area of the ^{31}P NMR spectra, or as concentration (mg P kg^{-1}soil; value in parenthesis). The latter was calculated by multiplying NaOH-EDTA extract P by the relevant percentage of peak area. Data are adapted from Dou et al. (2009)

Soil Series	Manure	Extraction Recovery	Inorganic			Organic		
			OrthoP	PyroP	PolyP	Monoesters		Diesters
						phytate	Other	
		(% of total P)						
Bucks	Poultry	84.2	90.4 (1919)	0.54 (12)	1.25 (27)	3.24 (68)	4.60 (98)	0
	Unt [a]	73.7	70.7 (447)	1.09 (7)	0	6.83 (43)	21.40 (135)	0
Chester	Dairy	79.1	88.7 (1958)	0.43 (9)	2.15 (47)	3.80 (84)	4.73 (104)	0.15 (3)
	Unt	84.4	60.7 (443)	1.35 (10)	0	18.34 (134)	18.11 (132)	1.53 (11)
Duffield 1	Dairy	80.4	89.2 (1617)	0.44 (8)	0.48 (9)	4.45 (81)	5.47 (99)	0
	Unt	68.4	59.4 (278)	2.37 (11)	0	11.68 (55)	26.58 (124)	0
Glenelg	Mushroom compost	77.9	45.0 (1707)	4.75 (180)	0	n/a [b]	47.24 (1790)	2.88 (109)
	Unt	68.0	46.1 (189)	4.53 (19)	0	19.42 (80)	29.98 (123)	0
Watson	Swine	69.6	80.9 (640)	1.33 (11)	0	8.41 (66)	9.40 (74)	0
	Unt	97.4	34.3 (213)	4.85 (30)	0	21.95 (136)	36.35 (226)	2.55 (16)
Duffield 2	Poultry	91.2	88.7 (1597)	0.08 (1)	0.71 (13)	4.78 (86)	5.70 (103)	0
Hagerstown	Poultry	77.9	81.3 (1055)	0.69 (9)	0.71 (9)	7.30 (95)	9.66 (125)	0.36 (5)
Manor	Dairy	72.9	79.4 (467)	1.25 (7)	n/a	n/a	19.40 (114)	n/a
Berks/Weikert 1	Duck	79.9	93.0 (2686)	0.35 (10)	1.51 (44)	1.81 (52)	5.1 (147)	0
Berks/Weikert 2	Poultry	82.7	89.9 (2571)	0.36 (10)	1.24 (35)	4.07 (117)	4.43 (127)	0

[a]: Receiving no manure or fertilizer.

[b]: Not available due to poor resolution.

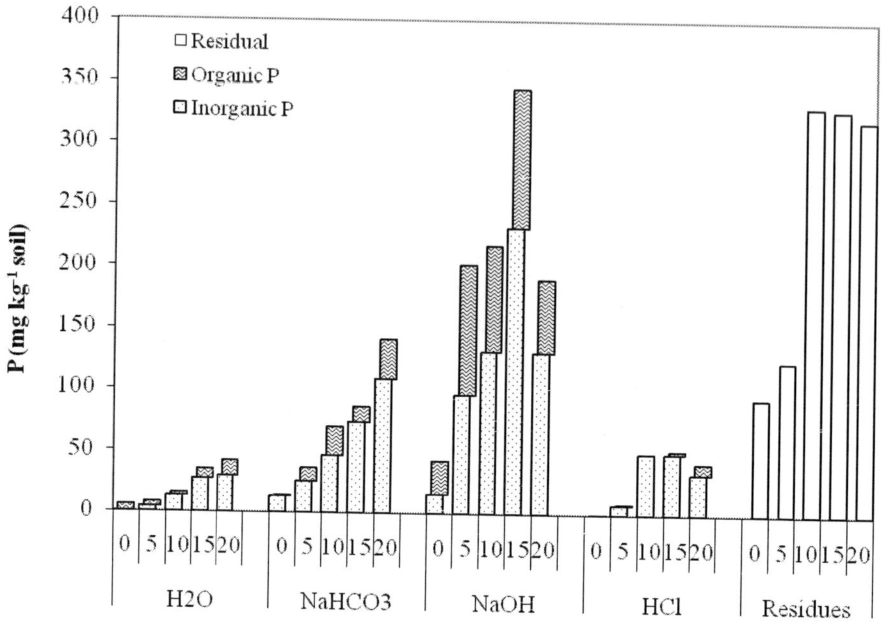

Figure 4. Phosphorus distribution in sequentially extracted H_2O, 0.5 M $NaHCO_3$ (pH8.5), 0.1 M NaOH, 1 MHCl and residual fraction of the surface (0-20 cm) soils collected from the five fields with none (0), 5-, 10-, 15-, and 20-yr of poultry litter application (He et al. unpublished data)

Repeated PL application also increased P in the subsurface horizons (20-40, and 40-60 cm) but at a lower accumulation rate (data not shown). The concentrations of H_2O-extractable P in the 40-60 cm soils after 15- and 20-yr PL application were one-third to one-quarter less than corresponding H_2O-extractable P levels in the surface 0-20 cm soils, indicating there had been little P leaching after long-term PL application in these soils. Although the distribution patterns of P in the two subsurface soils were similar to that of surface soils, the H_2O, $NaHCO_3$ and NaOH fractions of the subsurface soils, especially in the 40-60 cm layer, contained more P_o than their counterpart fractions in the surface soil. Similar to the surface soil, the levels of HCl-extractable P in the two subsurface layers did not correlate to any parameter of the PL applications.

These results not only reveal the transformation mechanism of PL P in pasture soils, but also have implications for environmental impacts. These data clearly indicate that runoff potential (i. e. the level of water extractable P) is more closely related to years of PL application than the annual application rate over the long term. A model incorporating the number of years of PL application could improve the prediction of surface P runoff from long-term PL applied fields (Vadas et al., 2007), thus providing insights to develop soil management and technology for efficiently reducing runoff (or erosion) rate to reverse surface water eutrophication.

REFERENCES

Abbasi, M. K., Hina, M., Khalique, A. & Khan, S. R. (2007). Mineralization of three organic manures used as nitrogen source in a soil incubated under laboratory conditions. Commun. *Soil Sci. Plant Anal*, *38*, 1691-1711.

Ajiboye, B., Akinremi, O. O. & Racz, G. J. (2004). Laboratory characterization of phosphorus in fresh and oven-dried organic amendments. *J. Environ. Qual*, *33*, 1062-1069.

Ajiboye, B., Akinremi, O. O., Hu, Y. & Flaten, D. N. (2007). Phosphorus speciation of sequential extracts of organic amendments using nuclear magnetic resonance and X-ray absorption near-edge structure spectroscopies. *J. Environ. Qual*, *36*, 1563-1576.

Barnett, G. M. (1994a). Phosphorus forms in animal manure. *Bioresour. Technol*, *49*, 139-147.

Barnett, G. M. (1994b). Manure P fractionation. Bioresour. *Technol*, *49*, 149-155.

Beauchemin, S., Hesterberg, D., Chou, J., Beauchemin, M., Simard, R. R. & Sayers, D. E. (2003). Speciation of phosphorus in phosphorus-enriched agricultural soils using X-ray absorption near-edge structure spectroscopy and chemical fractionation. *J. Environ. Qual*, *32*, 1809-1819.

Bowman, R. A. & Moir, J. O. (1993). Basic EDTA as an extractant for soil organic phosphorus. *Soil Sci. Soc. Am. J*, *57*, 1516-1518.

Buehler, S., Oberson, A., Rao, I. M., Friesen, D. K., & Frossard, E. (2002). Sequential phosphorus extraction of a ^{33}P-labeled oxisol under contrasting agricultural systems. *Soil Sci. Soc. Am. J*, *66*, 868-877.

Cade-Menun, B. J. (2005a). Using phosphorus-31 nuclear magnetic resonance spectroscopy to characterize phosphorus in environmental samples., 21-44, *In* B. L. Turner, et al., eds. *Organic Phosphorus in the Environment*. CABI Publishing, Cambridge, MA.

Cade-Menun, B. J. (2005b). Characterizing phosphorus in environmental and agfricultural samples by ^{31}P nuclear magentic resonance spectroscopy. *Talanta*, *66*, 359-371.

Codling, E. E. (2006). Laboratory characterization of extractable phosphorus in poultry litter and poultry litter ash. *Soil Sci.*, *171*, 858-864.

Cross, A. F. & Schlesinger, W. H. (1995). A literature review and evaluation of the Hedley fractionation: applications to the biogeochemical cycle of soil phosphorus in natural ecosystems. *Geoderma*, *64*, 197-214.

Dail, H. W., He, Z., Erich, M. S. & Honeycutt, C. W. (2007). Effect of drying on phosphorus distribution in poultry manure. Commun. *Soil Sci. Plant Anal*, *38*, 1879-1895.

Dao, T. H. & Hoang, K. Q. (2008). Dephosphorylation and quantification of organic phosphorus in poulry litter by purified phytic-acid high affinity *Aspergillus* phosphohydrolases. *Chemosphere*, *72*, 1782-1787.

Dao, T. H., Lugo-Ospina, A., Reeves III, J. B. & Zhang, H. (2006). Wastewater chemsitry and fractionation of bioactive phosphorus in dairy manure. Commun. *Soil Sci. Plant Anal*, *37*, 907-924.

Dick, W. A. & Tabatabai, M. A. (1977). Determination of orthophosphate in aqueous solutions containing labile organic and inorganic phosphorus compounds. *J. Environ. Qual*, *6*, 82-85.

Dou, Z., Toth, J. D., Galligan, D. T., Ramberg, C. F. Jr., & Ferguson, J. D. (2000). Laboratory procedures for characterizing manure phosphorus. *J. Environ. Qual*, *29*, 508-514.

Dou, Z., Ramberg, C. F., Toth, J. D., Wang, Y., Sharpley, A. N., Boyd, S. E., Chen, C. R., Williams, D. & Xu, Z. H. (2009). Phosphorus speciation and sorption-desorption characteristics in heavily manured soils. *Soil Sci. Soc. Am. J*, *73*, 93-101.

Dou, Z., Knowlton, K. F., Kohn, R. A., Wu, Z., Satter, L. D., Zhang, G., Toth, J. D. & Ferguson, J. D. (2002). Phosphorus characteristics of dairy feces affected by diets. *J. Environ. Qual*, *31*, 2058-2065.

Eghball, B., Wienhold, B. J., Woodbury, B. L. & Eigenberg, R. A. (2005). Plant availability of phosphorus in swine slurry and cattle feedlot manure. *Agron. J*, *97*, 542-548.

Griffin, T. S., He, Z. & Honeycutt, C. W. (2005). Manure composition effects net transformations of nitrogen from dairy manures. *Plant Soil*, *273*, 29-38.

Gungor, K., Jurgensen, A. & Karthikeyan, K. G. (2007). Determination of phosphorus speciation in dairy manure using XRD and XANES spectroscopy. *J. Environ. Qual*, *36*, 1856-1863.

Haygarth, P. M. & Sharpley, A. N. (2000). Terminology for phosphorus transfer. *J. Environ. Qual*, *29*, 10-15.

He, Z. & Honeycutt, C. W. (2001). Enzymatic characterization of organic phosphorus in animal manure. *J. Environ. Qual*, *30*, 1685-1692.

He, Z. & Honeycutt, C. W. (2005). A modified molybdate blue method for orthophosphate determination suitable for investigating enzymatic hydrolysis of organic phosphates. Commun. Soil Sci. *Plant Anal*, *36*, 1373-1383.

He, Z., Honeycutt, C. W. & Griffin, T. S. (2003a). Comparative investigation of sequentially extracted P fractions in a sandy loam soil and a swine manure. Commun. *Soil Sci. Plant Anal*, *34*, 1729-1742.

He, Z., Honeycutt, C. W. & Griffin, T. S. (2003b). Enzymatic hydrolysis of organic phosphorus in extracts and resuspensions of swine manure and cattle manure. *Biol. Fertil. Soils*, *38*, 78-83.

He, Z., Griffin, T. S. & Honeycutt, C. W. (2004a). Enzymatic hydrolysis of organic phosphorus in swine manure and soil. *J. Environ. Qual*, *33*, 367-372.

He, Z., Griffin, T. S. & Honeycutt, C. W. (2004b). Phosphorus distribution in dairy manures. *J. Environ. Qual*, *33*, 1528-1534.

He, Z., Griffin, T. S. & Honeycutt, C. W. (2004c). Evaluation of soil phosphorus transformations by sequential fractionation and phosphatase hydrolysis. *Soil Sci.*, *169*, 515-527.

He, Z., Dao, T. H. & Honeycutt, C. W. (2006a). Insoluble Fe-associated inorganic and organic phosphates in animal manure and soil. *Soil Sci.*, *171*, 117-126.

He, Z., Griffin, T. S. & Honeycutt, C. W. (2006b). Soil phosphorus dynamics in response to dairy manure and inorganic fertilizer applications. *Soil Sci.*, *171*, 598-609.

He, Z., Senwo, Z. N., Mankolo, R. N. &. Honeycutt, C. W. (2006c). Phosphorus fractions in poultry litter characterized by sequential fractionation coupled with phosphatase hydrolysis. *J. Food Agri. Environ.*, *4*, 304-312.

He, Z., Toor, G. S., Honeycutt, C. W. & Sims, J. T. (2006d). An enzymatic hydrolysis approach for characterizing labile phosphorus forms in dairy manure under mild assay conditions. *Bioresour. Technol.*, *97*, 1660-1668.

He, Z., Honeycutt, C. W., Zhang, T. & Bertsch, P. M. (2006e). Preparation and FT-IR characterization of metal phytate compounds. *J. Environ. Qual*, *35*, 1319-1328.

He, Z., Tazisong, I. A., Senwo, Z. N. & Zhang, D. (2008a). Soil properties and macro cations status impacted by long-term applied poultry litter. Commun. *Soil Sci. Plant Anal, 39*, 858-872.

He, Z., Honeycutt, C. W., Zhang, T., Pellechia, P. J. & Caliebe W. A. (2007a). Distinction of metal species of phytate by solid state spectroscopic techniques. *Soil Sci. Soc. Am. J, 71*, 940-943.

He, Z., Honeycutt, C. W., Cade-Menun, B. J., Senwo, Z. N. & Tazisong, I. A. (2008b). Phosphorus in poultry litter and soil, enzymatic and nuclear magnetic resonance characterization. *Soil Sci. Soc. Am. J, 72*, 1425-1433.

He, Z., Waldrip, H. W., Erich, M. S. Honeycutt, C. W. & Senwo, Z. N. (2009a). Enzymatic quantification of phytate in animal manure. Commun. *Soil Sci. Plant Anal, 40*, 566-575.

He, Z., Tazisong, I. A., Senwo, Z. N., Honeycutt, C. W. & Zhang, D. (2009b). Nitrogen and phosphorus accumulation in pasture soil from repeated poultry litter application. Commun. *Soil Sci. Plant Anal, 40*, 587-599.

He, Z., Honeycutt, C. W., Ohno, T., Wu, F. & Zhang, R. (2009c). Characterization of phosphorus associated with natural organic matter., p. 265-279, *In* F. Wu, & B. Xing, eds. *Natural Organic Matter and its Significance in the Environment*. Science Press, Beijing, China.

He, Z., Fortuna, A., Senwo, Z. N., Tazisong, I. A., Honeycutt, C. W. & Griffin, T. S. (2006f). Hydrochloric fractions in Hedley fractionation may contain inorganic and organic phosphates. *Soil Sci. Soc. Am. J, 70*, 893-899.

He, Z., Honeycutt, C. W., Xing, B., McDowell, R. W., Pellechia, P. J. & Zhang, T. (2007b). Solid-state Fourier transform infrared and ^{31}P nuclear magnetic resonance spectral features of phosphate compounds. *Soil Sci., 172*, 501-515.

He, Z., Cade-Menun, B. J., Toor, G. S., Fortuna, A., Honeycutt, C. W. & Sims, J. T. (2007c). Comparison of phosphorus forms in wet and dried animal manures by solution phosphorus-31 nuclear magnetic resonance spectroscopy and enzymatic hydrolysis. *J. Environ. Qual, 36*, 1086-1095.

He, Z., Honeycutt, C. W., Griffin, T. S., Cade-Menun, B. J., Pellechia, P. J. & Dou, Z. (2009d). Phosphorus forms in conventional and organic dairy manure identified by solution and solid state P-31 NMR spectroscopy. *J. Environ. Qual, 38*, 1909-1918.

Hedley, M. J., Stewart, J. W. B. & Chauhan, B. S. (1982a). Changes in inorganic and organic soil phosphorus fractions induced by cultivation practices and by laboratory incubations. *Soil. Sci. Soc. Am. J, 46*, 970-976.

Hedley, M. J., White, R. E. & Nye, P. H. (1982b). Plant-induced changes in the rhizosphere of rape (*Brassica napus* var.Emerad) seedlings: III. Changes in L value, soil phosphate fractions and phosphatase activity. *New Phytol*, *91*, 45-56.

Hill, J. E. & Cade-Menun, B. J. (2009). Phosphorus-31 nuclear magnetic resonance spectroscopy transect study of poultry opearations on the Delmarva Peninsula. *J. Environ. Qual*, *38*, 130-138.

Hunger, S., Sims, J. T. & Sparks, D. L. (2005). How accurate is the assessment of phosphorus pools in poultry litter by sequential extraction? *J. Environ. Qual*, *34*, 382-389.

Hunger, S., Cho, H., Sims, J. T. &. Sparks, D. L. (2004). Direct speciation of phosphorus in alum-amended poultry litter: solid-state ^{31}P NMR investigation. *Environ. Sci. Technol.*, *38*, 674-681.

Jayasundera, S., Schmidt, W. F., Reeves III, J. B. & Dao, T. H. (2005). Direct ^{31}P NMR spectroscopic measurement of phosphorus forms in dairy manures. *J. Food Agri. Environ.*, *3*, 328-333.

Kleinman, P., Sullivan, D., Wolf, A. M., Brandt, R., Dou, Z., Elliott, H., Kovar, J., Leytem, A., Maguire, R. O., Moore Jr., P. A., Saporito, L., Sharpley, A. N., Shober, A. L., Sims, J. T., Toth, J. D., Toor, G. S. Zhang, H. & Zhang, T. (2007). Selection of a water-extractable phosphorus test for manures and biosolids as an indicator of runoff loss potential. *J. Environ. Qual*, *36*, 1357-1367.

Kleinman, P. J. A., Sharpley, A. N., Wolf, A. M., Beegle, D. B. & Moore, P. A. Jr. (2002). Measuring water-extractable phosphorus in manure as an indicator of phosphorus in runoff. *Soil Sci. Soc. Am. J*, *66*, 2009-2015.

Kruse, J., Leinweber, P., Eckhardt, K., Godlinski, F., Hu, Y. & Zuin, L. (2009). Phosphorus $L_{2,3}$-edge XANES: overview of reference compounds. *J. Synchrotron Rad*, *16*, 247-259.

Leinweber, P., Haumaier, L. & Zech, W. (1997). Sequential extractions and ^{31}P-NMR spectroscopy of phosphorus forms in animal manures, whole soils and particle-size separates from a densely populated livestock area in northwest Germany. *Biol. Fertil. Soils*, *25*, 89-94.

Leytem, A. & Thacker, P. A. (2008). Fecal phosphorus excretion and characterization from swine fed diets containing a variety of cereal grains. *J. Anim. Veterian. Adv*, *7*, 113-120.

Leytem, A. B., Turner, B. L. & Thacker, P. A. (2004). Phosphorus composition of manure from swine fed low-phytate grains: Evidence for hydrolysis in the animal. *J. Environ. Qual*, *33*, 2380-2383.

Leytem, A. B., Smith, D. R., Applegate, T. J. & Thacker, P. A. (2006). The influence of manure phytic acid on phosphorus solubility in calcareous soils. *Soil Sci. Soc. Am. J*, *70*, 1629-1638.

Maguire, R. O., Sims, J. T., Saylor, W. W., Turner, B. L., Angel, R. & Applegate, T. J. (2004). Influence of phytase addition to poultry diets on phosphorus forms and solubility in litters and amended soils. *J. Environ. Qual*, *33*, 2306-2316.

Makarov, M. I. (2005). Phosphorus-containing compounds of soil organic matter: ^{31}P NMR spectroscopic study (a review). *Eurasian Soil Sci.*, *38*, 153-164.

McAuliffe, C. & Peech, M. (1949). Utilization by plants of phosphorus in farm manure: *I. Soil Sci.*, *68*, 179-184.

McDowell, R. W. & Stewart, I. (2005). Phosphorus in fresh and dry dung of grazing dairy cattle, deer, and sheep: Sequential fraction and phosphorus-31 nuclear magnetic resonance analysis. *J. Environ. Qual*, *34*, 598-607.

Murphy, J. & Riley, J. P. (1962). A modified single solution method for the determination of phosphate in natural waters. *Analytica Chimia Acta*, *27*, 31-36.

Pant, H. K., Edwards, A. C. & Vaughan, D. (1994). Extraction, molecular fractionation and enzyme degradation of organically associated phosphorus in soil solutions. *Biol. Fertil. Soils*, *17*, 196-200.

Peak, D., Sims, J. T. & Sparks, D. L. (2002). Solid-state speciation of natural and alum-amended poultry litter using XANES spectroscopy. *Environ. Sci. Technol*, *36*, 4253-4261.

Peperzak, P., Caldwell, A. G., Hunziker, R. R. & Black, C. A. (1959). Phosphorus fractions in manures. *Soil Sci.*, *87*, 293-302.

Preusch, P. L., Adler, P. R., Sikora, L. J. & Tworkoski, T. J. (2002). Nitrogen and phosphorus availability in composted and uncomposted poultry litter. *J. Environ. Qual*, *31*, 2051-2057.

Qian, P. & Schoenau, J. J. (2000). Fractionation of P in soil as influenced by a single addition of liquid swine manure. *Can. J. Soil Sci.*, *80*, 561-566.

Sato, S., Solomon, D., Hyland, C., Ketterings, Q. M. & Lehmann, J. (2005). Phosphorus speciation in manure and manure-amended soils using XANES spectroscopy. *Environ. Sci. Technol.*, *39*, 7485-7491.

Shand, C. A., Coutts, G., Hillier, S., Lumsdon, D. G., Chudek, A. & Eubeler, J. (2005). Phosphorus composition of sheep feces and changes in the field determined by ^{31}P NMR spectroscopy and XRPD. *Environ. Sci. Technol*, *39*, 9205-9210.

Sharpley, A. N. & Moyer, B. (2000). Phosphorus forms in manure and compost and their release during simulated rainfall. *J. Environ. Qual*, *29*, 1462-1469.

Sharpley, A. N., McDowell, R. W. & Kleinman, P. J. A. (2004). Amounts, forms, and solubility of phosphorus in soils receving manure. *Soil Sci. Soc. Am. J*, *68*, 2048-2057.

Shober, A. L., Hesterberg, D. L., Sims, J. T. & Gardner, S. (2006). Characterization of phosphorus species in biosolids and manures using XANES spectroscopy. *J. Environ. Qual*, *35*, 1983-1993.

Toor, G. S., Condron, L. M., Di, H. J., Cameron, K. C. & Cade-Menun, B. J. (2003). Characterization of organic phosphorus in leachate from a grassland soil. *Soil Biol. Biochem*, *35*, 1317-1323.

Toor, G. S., Peak, J. D. & Sims, J. T. (2005). Phosphorus speciation in broiler litter and turkey manure produced from modified diets. *J. Environ. Qual*, *34*, 687-697.

Turner, B. L. (2004). Optimizing phosphorus characterization in animal manures by solution phosphorus-31 nuclear magnetic resonance spectroscopy. *J. Environ. Qual*, *33*, 757-766.

Turner, B. L. & Leytem, A. B. (2004). Phosphorus compounds in sequential extracts of animal manure: chemical speciation and a novel fractionation procedure. *Environ. Sci. Technol.*, *38*, 6101-6108.

Turner, B. L., Mahieu, N. & Condron, L. M. (2003). Phosphorus-31 nuclear magnetic resonance spectral assignments of phosphorus compounds in soil NaOH-EDTA extracts. *Soil Sci. Soc. Am. J*, *67*, 497-510.

Uusitalo, R., Ylivainio, K., Turtola, E. & Kangas, A. (2007). Accumulation and translocation of sparingly soluble manure phosphorus in different types of soils after long-term excessive inputs. *Agric. Food Sci.*, *16*, 317-331.

Vadas, P. A. & Kleinman, P. (2006). Effect of methodology in estimating and interpreting water-extractable phosphorus in animal manures. *J. Environ. Qual*, *35*, 1151-1159.

Vadas, P. A., Gburek, W. J., Sharpley, A. N. Kleinman, P. J. A., Moore Jr., P. A., Cabrera, M. L. & Harmel, R. D. (2007). A model for phosphorus transformation and runoff loss for surface-applied manure. *J. Environ. Qual*, *36*, 324-332.

Waldrip-Dail, H., He, Z., Erich, M. S. & Honeycutt, C. W. (2009). Soil phosphorus dynamics in response to poultry manure amendment. *Soil Sci.*, *174*, 195-201.

Wienhold, B. J. & Miller. P. S. (2004). Phosphorus fractionation in manure from swine fed traditional and low-phytate corn diets. *J. Environ. Qual*, *33*, 389-393.

Wodzinski, R. J. & Ullah, A. H. (1996). Phytase. *Adv. Appl. Microbiol*, *42*, 263-302.

Ylivainio, K., Uusitalo, R. & Turtola, E. (2008). Meat bone meal and fox manure as P sources for ryegrass (*Lolium multiflorun*) grown on a limed soil. Nutr. *Cycl. Agroecosyst*, *81*, 267-278.

Reviewed by Barbara J. Cade-Menun, Agriculture and Agri-Food Canada; Peter A. Vadas, USDA-ARS Dairy Forage Research Center; and Hailin Zhang, Dept. of Plant and Soil Sciences, Oklahoma State University.

In: Manure: Management, Uses and… ISBN: 978-1-61668-424-2
Editor: C. S. Dellaguardia, pp. 119-144

Chapter 4

IMPROVING PERFORMANCE AND MANURE MANAGEMENT IN THE FRENCH PIG SECTOR: A THREE-STAGE ANALYSIS

Isabelle Piot-Lepetit
European Commission, Joint Research Centre (JRC),
Institute for Prospective Technological Studies (IPTS), Seville, Spain
and INRA, UMR 1302 SMART, F-35000, Rennes, France[1]

ABSTRACT

Nitrogen emissions from agriculture are considered as an important problem, motivating an evaluation of the performance of French pig farms in improving manure management. This article develops a methodology that allows for a joint evaluation of farms' economic and environmental efficiency from various productive structures. A three-step approach is developed. First, a Cluster Analysis is implemented. It provides a description of main French pig farm structures. Farms are grouped into clusters with similar structural characteristics regarding land, labour and production diversification versus specialisation. Second, a Data Envelopment Analysis is implemented at the level of each cluster. This nonparametric frontier analysis allows for efficiency benchmarking. It provides a score on the economic performance of farms and on their

[1] The author would like to thank M. Le Moing, G. Jégouzo and M. Ulvé for their support in the implementation of statistical procedures developing in the paper.

efficiency in manure management in a way that reduces the detrimental impact of their productive activity on the environment. At last, an aggregate procedure is developed to measure the performance at the cluster level and at the sector level. This bottom-up approach allows us to provide an overall indicator of French pig farms' performance in both an economic and environmental direction and to identify the percentage of this performance that depends on farms' characteristics or on the structure of the pig sector.

Main results show that improvements both in economic and environmental terms are limited for farming systems specialized in pig production without any change in their production structure or technology of production. Farming systems that are diversified can contribute to an increase in production and a decrease in nitrogen surplus by implementing their current production technology in a most efficient way.

INTRODUCTION

One of the main concerns with the pig sector is the environmental impact of this production activity. Pigs are extremely efficient at converting feed to meat. However, the environmental consequences of pig production regarding the management of pig manure in relation to water and air pollution are of increasing public concern.

Pig production has historically been associated with other agricultural activities that provide livestock feed. Over time, pig production has become more regionally concentrated and more specialized relying on bought-in feed. The intense competition in the meat market and the decline in real prices received by farmers have been a favourable incitation for improvements in production, breeding and management techniques. Indeed, the competition on the world market induces pressure to lower production costs and encourages further intensification of production. These structural changes have enabled considerable productivity gains with an incentive to increase scale. The production has moved toward off-farm feed preparation. However, the intensification and regional concentration of pig production have raised the environmental degradation associated with this productive activity. A greater number of animals per farm leads to a larger volume of manure that must be disposed of. The reduction of land available per pig has induced an increase in the quantity of nutrients supplied to the soil and the potential harm to the environment.

The main environmental issues associated with pig production concern water and air pollution. Water pollution arises from the inappropriate disposal of pig manure. Nutrients in manure, principally nitrogen and phosphorus, are a significant component of pollution from agriculture to surface water, groundwater and marine waters, damaging ecosystems through eutrophication and degrading their recreational use. Water bodies can also be affected by organic effluents and pathogens contained in manure. In some areas, the emission of ammonia from livestock housing facilities and from badly managed storage and spreading of manure are of serious concern. Pig production can also be a source of greenhouse gas emissions, mainly methane and nitrous oxide, but its importance is smaller. At the locale, odour can be a serious air pollution problem resulting from pig production (OECD, 2003).

In terms of policy measures, the response by the European Union (EU) to address environmental issues in the livestock sector was to impose regulations through the directive concerning the protection of waters against pollution caused by nitrates from agriculture sources (hereafter called the Nitrate Directive)[2]. This directive implied three main requirements. Firstly, the EU member states had to identify areas where water quality is threatened by nitrate pollution from agriculture. These land areas where agricultural production contributes to drinking water problems are designated as Nitrate Vulnerable Zones (NVZs). Secondly, member states had to establish codes of good agricultural practices in order to improve the protection of water bodies against nitrate pollution. They include rules relating to periods when the land application of certain types of fertilizers are prohibited, minimum capacity for manure storage facilities and limitations on the land application of fertilizers consistent with good agricultural practices in each vulnerable zone. Thirdly, member states had developed action programmes for their NVZs. These mandatory programmes consist of the standards contained in the voluntary codes with the objective of complying with the limit on the application of animal manure specifically targeted in the Nitrate Directive, which sets a maximum rate of 170 kg of nitrate per hectare.

Technologies such as housing holding facilities, manure storage and treatment systems, and management techniques such as altering feed composition and manure spreading practices aimed at lowering the environmental costs associated with pig production have been developed. Some of these are economically beneficial to producers, while others require significant financial investments or an increase in variable costs. The feeding

[2] Directive 91/676/EEC

practices, manure spreading management and farming systems used in rearing pigs have important implications for the environmental outcomes from pig production. Changes in pig feed composition or increased feed conversion efficiency can lead to a reduction in nutrient excretions per unit of production. The choice of management practice to spread pig slurry on fields can considerably alter ammonia emission levels and nutrient soil surface run-off and leaching. The more efficient method (soil injection) of manure application in fields is the most costly practice. Thus, polluting averting technologies are not considered to be scale neutral. Large-scale farming systems have a greater potential to introduce such technologies because the cost can be spread over a larger volume of production. Moreover, these large operations can take advantage of the "state of the art" technologies in controlling pollution, which smaller producers are unable to afford or manage (Norris and Thurow, 1999). From another side, regulation could give significant cost advantages to land extensive mixed farms integrating crop and pig production.

While different types and scale of production have potentially different effects on the environment, little evidence is available to judge between systems. Comparing the efficiency of different pig farming systems in controlling environmental pollution is complex. This is because of the large array of pig production systems, ranging from indoor to outdoor systems, small to large fully integrated production operations, extensive to intensive units, through to organic rearing of pigs. Also while one particular system might be highly efficient in producing pig meat in terms of economic cost, it might be poorer in attaining high standards in terms of environmental objectives or vice versa. Furthermore, the performance of nutrient emission can also vary considerably between farms using the same pig husbandry practices (Backus et al., 1998).

At present, there is little empirical work to evaluate the relative efficiency in economic and environmental terms of different pig rearing systems and scale of production. This paper is an attempt to cover this point. The aim of the paper is to provide an evaluation of the efficiency of farms from the French pig sector in a way that increase their economic performance and also improve their manure management. A three-stage approach is developed.

The first stage uses statistical techniques to derive groups of farms that share a common production structure. The Cluster analysis is a procedure that is used to capture the natural structure of data sets by dividing them into classes that are meaningful groups. This approach is implemented in this paper to provide a classification of the French farms with pig farming activities, built on three main items: land use, labour use and diversity of production versus specialisation in pig production. The extraction of this structural information from the data set allows us to define clusters of farms with a homogenous productive structure. This step is of importance for the evaluation of performance since it will avoid comparing disparate firms and searching for relationships among heterogeneous production units.

Then, a nonparametric frontier analysis, called Data Envelopment Analysis (DEA), which allows efficiency measurement is used. Built upon the previous works of Farrell (1957) and Charnes et al. (1978), DEA is a non parametric methodology used to assess relative efficiencies of a set of comparable Decision Making Units (DMUs), here farms sharing the same production technology within a cluster. DEA is a "data oriented" approach. The definition of a DMU is generic and quite flexible (Cooper et al., 2004). The literature provides a large variety of applications of the DEA methodology for estimating the performance of many different kinds of DMUs, having operations in many industries, in many contexts and in many countries (Seiford, 1997; Gattoufi et al., 2004; Emrouznejad et al., 2008). The measurement of efficiency is accomplished without explicitly requiring expressed assumptions and has the advantage of avoiding the need for assigning a priori measures of relative importance to any input or output (Seiford and Thrall, 1990). Furthermore, it avoids the need for explicitly specifying the formal relationships that are supposed to exist between inputs and outputs. Because it requires very few assumptions, DEA has also been used in cases which have been resistant to other methodological approaches owing to the complex nature of the relationships between the multiple inputs and the multiple outputs involved in DMUs. This methodology also avoids the need for having recourse to prices or other assumptions on weights which are supposed to reflect the relative importance of the different inputs or outputs. DEA is a methodology directed to frontier rather than central tendencies. Instead of trying to fit a regression plane, through the centre of the data, as in statistical regression models, DEA puts a piecewise linear surface[3] on top of the observations (Cooper et al., 2004). In doing this, DEA is proven particularly useful at uncovering relationships that remain hidden from other methodologies.

Recently, the DEA method has been extended to integrate undesirable outputs or pollutants in the modelling. Among others, Ball et al. (1994), Färe et al. (1996) or Tyteca (1997) adapted DEA to construct environmental efficiency measurements. These various measures depend on how undesirable

[3] DEA models are specified as linear programming models.

output reductions are sought. However, in all these studies, detrimental outputs are considered as weakly disposable. This hypothesis restricts potential adjustments of desirable goods and is considered as a transcription of the cost induced by the environmental regulation on production. In this paper, the undesirable output from pig production is the nitrogen surplus. However, its disposal in the environmental is not free in the sense that it induces degradation costly to repair. Thus, nitrogen surplus is introduced in the DEA model as a weakly disposable output. Furthermore, the evaluation of farm's economic and environmental performance is done by using a directional distance function, as defined by Chung et al. (1997). This efficiency measure evaluates performance in a direction which allows for an increase in production only with a joint decrease in nitrogen surplus. The use of this directional distance function provides an evaluation of improvements in efficiency of pig farms that can be reached without increasing environmental damages.

Finally, an aggregation procedure is used. The purpose is to develop knowledge about the implication of economic and environmental efficiency of farms with a pig farming activity at the sector level. As there is various way of producing pig in France, our approach accounts for this diversity. First, a measure of the efficiency at the cluster level is provided that integrate individual variability. This measure summarizes the possible improvement that could be reached if all farms in the cluster become efficient by increasing its production level and decreasing its manure surplus. Then, a measure of the efficiency of the sector as a whole is defined by taking into account the heterogeneity existing between pig farming systems. The measure provides information on improvements that could be expected at the sector level if each cluster becomes efficient. It also allows for a characterization of farming systems that outperform in the joint management of economic and environmental performance.

The paper is organised as follows. Section 2 describes the three-stage analysis implemented for identifying the diversity and specificity of pig farming systems in France and evaluating their economic and environmental efficiency at the farming system level and at the sector level as a whole. Data and the selection of variables are presented in Section 3. Section 4 deals with main findings from this paper. Finally, section 5 concludes.

METHODOLOGY

The methodology is a three-stage analysis. First, a cluster analysis allows us to classify farms with similar structural and productive characteristics in a common cluster. Several statistical tools are used for validating this classification over time. Second, a methodology called Data Envelopment Analysis is implemented for measuring the efficiency of farms in increasing their production and jointly decreasing their environmental detrimental impact. The third stage is based on an aggregation procedure allowing for a measurement of efficiency at the sector level. The cluster efficiency measure characterizes the variability in efficiency within a farming system due to individual farms' heterogeneity. The sector efficiency measure is linked to the structure of the production sector and identifies farming systems that have the highest contribution to the sector efficiency.

First Stage: A Cluster Analysis

The objective of the cluster analysis is to find information from the data set to obtain a synthetic view of the main French pig farming systems. A cluster analysis is a generic term for a large number of methods which attempt to determine whether or not a data set contains distinct groups of observations and, if so, to find which observations belong together. Before using clustering techniques a principal component analysis was implemented with the objective to reduce all the variables available to a small set explaining a large portion of the original data set.

The most commonly used clustering methods is the *hierarchical classification* (HC), which lead to series of nested clusters of observations, beginning at the stage where each observation is considered as a separate group and ending with one group containing all observations. The series of grouping are summarized by mean of a tree, called dendrogram. Solutions for a particular number of clusters can be selected by cutting the dendrogram at specific levels. Hierarchical classification uses a distance matrix as starting point. The elements of this matrix give a measure of the distance between a pair of observations. An HC with the Centroïd Approach was first used. In this case the distance between clusters is defined as the Euclidean distance between the respective cluster means. This approach was not enough powerful for grouping farms since it defined some clusters with only one farm. Thus, another approach was used. An HC with the Ward's Minimum Variance Approach is a statistical tool based on the assumption that if two clusters are very similar and hence close to each other, then the between cluster sum of squares should be small.

To confirm classification provided by the hierarchical clustering procedures, a non hierarchical clustering method called *k-Means Clustering Approach* was also used. This approach is very effective for large data sets and can be summarized as follows. First, some seeds are selected to act as the initial centroïds for the k clusters. Then, each observation is assigned to the cluster with the nearest seed. The measure used is the Euclidian distance. When all observations have been assigned, the final clusters result.

Then, the validation of the classification by statistical methodologies is an important step for evaluating the stability of the results. Several tests have been implemented as a Kruskal-Wallis test and a Discriminant analysis.

The *Kruskal-Wallis test* which is a generalisation of the two-sample Wilcoxon test to several groups, analyses differences in central tendencies among data. This nonparametric test allows us to verify if variable distributions are the same in each class or not. If distributions of variable in two distinct classes are different then the two classes are expected to really differentiate between farms with dissimilar production structures.

The *Discriminant Analysis* focuses on separating distinct sets of observations and then allocating the observations into these sets. This approach evaluates the probability of misclassification of unknown observations in the classification defined by the cluster analysis. Three methods have been used for estimating misclassification rates. The first one is applied to derive classification rules in the original data (the training set) and to assess misclassification rate estimates on the same dataset. This approach is known as the *resubstitution method.* Estimating misclassification rates in this way is known to be generally optimistic. More realistic misclassification rate estimates are defined by using another resubstitution method, called *leaving one out method* in which the discriminant function is derived from the original data minus one observation and used to classify that observation. This process is repeated for each observation in the data set, each time omitting a different observation. The last method is the *cross-validation method.* It implies a separation of the original data set between a training set and a test set. The training set is used to derive discriminant rules while the test set is used to assess misclassification rates. This approach provides generally unbiased estimates of the true misclassification rates.

Stage Two: Farm's Efficiency Measurement with a Data Envelopment Analysis

The Data Envelopment Analysis (DEA)[4] is a nonparametric method in the sense that DEA is entirely based on observed inputs and outputs used by farmers in their production processes. Thus, DEA is a very flexible way of modelling production technologies. The choice of a given DEA model always depends on underlying economic and technological assumptions on production processes that transform inputs into outputs.

The objective of this second stage is the assessment of a measurement of the productive performance of different French pig farming systems. Thus, the performance is defined in an environmental preventive management framework in such a way that efficiency gains, which imply changes in production processes, are assessed when a producer can provide greater amounts of desirable outputs together with lower polluting emission levels. The DEA enables us to determine the success of a producer in attaining the objective of producing the largest increase in desirable outputs and decrease in undesirable outputs. Each producer receives a score, which results from the comparison of his productive bundle with those from other producers using the same technology. The efficiency score is unity when no famer develops better practices to fulfil the defined objective. Otherwise, the score is upper unity and gives an information on the potential productive gain that can be realised by a change in the implementation of the production technology in a way that have already been done by the most efficient producers.

From a theoretical point of view, we consider that the production of desirable goods such as pig meat is accompanied by the joint production of undesirable by-products such as manure surplus or nitrogen surplus. In a context of environmental regulation, we define a DEA model that improves efficiency through an increase in desirable outputs and a decrease in undesirable outputs (Chung et al., 1997). Furthermore, the reduction of undesirable by-products is assumed to be costly. Resources must be diverted to clean up, production levels must be cut back or fines must be paid. This implies a weak disposability assumption for polluting goods.

To be formal, let us consider J firms or DMUs ($j=1,\dots,J$) in a productive sector. We assume that the sector can be decomposed in K homogeneous groups of firms ($k=1,\dots,K$). Each firm produces good outputs $y_j \in R_+^M$,

[4] For more information on the origins of DEA, see Førsund and Sarafoglou (2002).

undesirable or bad outputs $b_j \in R_+^S$ by employing inputs $x_j \in R_+^N$. Then, the technology of each homogeneous group of firms can be described in a very general way by an output set:

$$P^k(x_j) = \{(y_j, b_j) : x_j \text{ can produce} (y_j, b_j)\}, j=1,...,J^k \quad (1)$$

where J^k is the number of firms j in group k ($J = \sum_{k=1}^{K} J^k$). In words, the output set $P^k(x_j)$ consists in all combinations of good and bad output vectors (y_j, b_j) that can be produced by the input vector x_j within the productive group k composed of J^k firms. All these J^k firms are assumed to share the same production technology defined by $P^k(x_j)$.

Following Färe et al. (1994), we model the idea that it is costly to reduce bad outputs by imposing an assumption of weak disposability on undesirable outputs as:

$$(y_j, b_j) \in P^k(x_j) \text{ et } 0 \leq \theta \leq 1 \text{ imply } (\theta y_j, b_j) \in P^k(x_j) \quad (2)$$

This assumption states that a reduction in undesirable outputs is feasible only if goods are simultaneously reduced, given a fixed level of inputs. This means that pollution abatement uses resources that, otherwise, could have been used to expand desirable production. In addition, we assume that the good outputs are freely disposable which means that they may be reduced without any cost.

$$(y_j, b_j) \in P^k(x_j) \text{ et } y_j^{'} \leq y_j \text{ imply } (y_j^{'}, b_j) \in P^k(x_j) \quad (3)$$

Then, the measurement of efficiency is defined in a way that allows for the possibility of crediting firms for the reduction in undesirable outputs and expansion in desirable outputs. A directional output distance function is used to represent technology, which is defined as (Chung et al., 1997; Färe and Grosskopf, 2004):

$$\vec{D}_o(x_j, y_j, b_j) = \sup\{\beta_j : (y_j, b_j) + \beta_j g \in P^k(x_j)\} \quad (4)$$

where g is the vector of direction in which outputs are scaled. In this paper, good and bad outputs are treated asymmetrically, thus $g=(y_j,-b_j)$. β_j is the maximum feasible expansion of the desirable outputs and contraction of the undesirable outputs in identical proportions for a given level of inputs. It is also the efficiency measure we are looking for. It is obtained by solving a linear programming model for each observation in the sample as defined in the appendix A1.

Let's note $E_j = 1+\beta_j$ the efficiency score of the farm j. If $E_j=1$ then the DMU j is on the best-practice frontier of its farming system. It is an efficient farm. This DMU could become a benchmark for inefficient DMUs in the same farming system. If $E_j>1$ then the DMU j lies below the best-practice frontier of its own group. It is an inefficient farm relatively to the performance of other farms within its farming system. This DMU can increase its production and decrease its level of pollution by implementing the production process of its farming system in a way done by DMUs located on its group's production frontier.

Stage Three: Sector Efficiency Measurement in Presence of Sample Heterogeneity

Providing efficiency measurement for a sector as a whole is complicated when there exits various ways for producing the same goods. For farms producing pig, the pig farming activity can be a diversification of their sources of revenues or the main production. To measure the efficiency at the sector level, we defined an approach that takes into account the multi-technologies structure of the French pig sector and the degree of efficiency of each farming system.

To provide these efficiency measures, we need to define two aggregates. The first one is the group's output that is defined as the sum of the individual output levels produced within the group, i.e.

$$y^k = \sum_{j=1}^{J^k} y_j \in R_+^M \text{, } k=1,\dots,K \qquad (5)$$

The second aggregate is the farming system's optimal output, which corresponds to the sum of the individual output levels corrected from their inefficiencies.

$$y^{k*}=\sum_{j=1}^{J^k}\beta_j y_j \in R_+^M \ , k=1,...,K \tag{6}$$

The comparison of these two aggregates allows us to provide a efficiency measurement for each group k ($k=1,...,K$) by using the following ratio:

$$\beta^k=\frac{y^{k*}}{y^k}=\frac{\sum_{j=1}^{J^k}\beta_j y_j}{\sum_{j=1}^{J^k}y_j}\ , k=1,...,K \tag{7}$$

The measure of the efficiency at the cluster level defined in (7) integrates individual heterogeneity. It summarizes the potential improvement that could be reached if all farms in the cluster become efficient by increasing their production level and decreasing their nitrogen surplus. This aggregation procedure is also used to measure efficiency at the sector level in the following way:

$$\beta=\frac{y^*}{y}=\frac{\sum_{k=1}^{K}y^{k*}}{\sum_{j=1}^{K}y^k}=\frac{\sum_{k=1}^{K}\beta^k y_k}{\sum_{k=1}^{K}y_k} \tag{8}$$

The β value is the sector efficiency measure. It is defined in a way that takes into account for the technological heterogeneity between each farming systems or groups of farms sharing a common production structure. The measure provides information on improvements that could be expected at the sector level if each cluster becomes efficient. It also allows for a characterization of farming sectors that outperform in the joint management of economic and environmental performance.

DATA

The data set is drawn from the French Farm Accountancy Data Network (FADN). The FADN is an annual survey of farms that is stratified according to region, economic size and type of farming. The selected data set contains all the farms having declared revenues from pig farming from 1996 to 2000. The sample contains around 3000 individuals (600 farms per year) after deletion for outliers. Furthermore, 187 variables were selected for describing farms' production structure.

Description of Variables Used in the Cluster Analysis

Variable selection is an important step for the cluster analysis to identify groups with a common productive structure. It should reflect the multiple organization dimension production structure of the sector. Experts have been involved in this task. They specified four dimensions that can characterise the structure of pig farming in France. The first one regards the use of land for both crops and livestock production. The second dimension considers working conditions on farm. The third one describes the intensity of farm's involvement in the pig rearing activity while the fourth dimension considers the diversification versus intensification of farm's activities.

After an analysis of correlation between the 187 available variables, only uncorrelated variables have been selected for the implementation of a Principal Component Analysis (PCA). The PCA is used to identify underlying constructs of factors that explain the correlation among the four dimensions defined by experts. It allows for a reduction of the variables in a small number of uncorrelated factors explaining most of the variability of the dataset. For selecting the right number of factors, three methods have been used. The first one consists in selecting all the factors having an eigenvalue upper that one. It results in a number of four factors. The second method is a selection of the factors corresponding to a percentage of the dataset variance upper that *1/NV%,* with *NV* the number of variables used in the PCA. Four factors are retained. The third one was a print of the histogram of the eigenvalues looking for an important fall in these values. The fourth first factors explained 70% of the total variance of the sample.

The first factor explains 33% of the variance of the sample. It separates farms with high land areas and a high herd of livestock from farms with few acreages of land. In the former group of farms, the herd is mainly composed of pig and cattle. The second factor explains 18% of the variance. It separates farms that have an intensive livestock production activity and an important pig herd from the others. The third factor explains 12% of the variance. It separates the diversified farms between those producing cattle and milk and those producing crops. The fourth factor explains only 6% the sample variance. It separates the intensified livestock farms producing only pig from those producing cattle, milk and pig.

Thus, the variable reduction resulted in four factors explaining 70% of the total variance. These four factors have been used for defining clusters of farms by using statistical techniques presented in the above section.

Description of DEA Variables

For the DEA approach, one good output, one undesirable output and four inputs have been defined for describing production processes.

The desirable output (y) is an aggregate of products sold by the farms in Euros 1996. It characterizes the potential revenue of each farm on the market resulting from its particular production structure.

Four inputs are used in the definition of production technologies:

Land (x_1) represents the total agricultural area of the farm in hectares,

Livestock herd (x_2) represents the size of the livestock herd on the farm in livestock units,

Labour (x_3) represents the total amount of family labour and hired labour used in annual worker units, and

Variable inputs (x_4) represent variable costs in Euros 1996.

We also consider the production of one undesirable output (b) which is an estimate of nitrogen surplus. This variable involves the calculation of individual farm nitrogen budgets, i.e., the balance between nitrogen inputs and outputs. Individual farm nitrogen budgets are constructed using estimates of flows of nitrogen. Annual quantities of nitrogen contained in purchased inputs (fertilizers and animal feeds) are summed and subtracted from the annual quantity of nitrogen contained in the harvested part of crops and in livestock sold off farms. Formally, this calculation is made as follows:

$$b = (N_{fert} + N_{feed} + N_{atmos}) - (N_{crop} + N_{lstk}) \qquad (9)$$

where b is the nitrogen surplus, N_{fert} the quantity of nitrogen contained in fertilisers, N_{feed} the quantity of nitrogen contained in animal feed, N_{atmos} the quantity of nitrogen deposited from the atmosphere, N_{crop} the quantity of nitrogen contained in harvested crops and N_{lstk} the quantity of nitrogen contained in livestock. All elements are measured in kilograms[5].

RESULTS

Cluster Analysis Results

The Cluster Analysis was implemented with two methods: the Hierarchical Analysis and the k-Means Clustering Analysis. The hierarchical analysis was implemented on 601 observations from year 1996 and with the 4 factors defined by the Principal Component Analysis. The cluster selection was obtained by using the Ward criterion, which groups observations implying the smallest variation of the interclass inertia. The selection of the partition of the sample into clusters was made using the cubic clustering criterion. This methodology suggested a typology in 4, 7, 10 or 12 clusters. The k-Means Clustering Analysis was implemented using an automatic detection of the first kernels of 60 clusters. Then only those with a number of observations upper than 10 were retained. It resulted in 16 initial centroids. After implementation of this method, a partitioning of the sample, based on the cubic clustering criterion, in 7, 10 or 12 classes was suggested. Then, the same methodology was reproduced for the each year from 1997 to 2000. Results were that a typology in 7 or 10 classes should provide a good representation of the underlying structure from the data set.

Finally, the final selection among these two classifications was the typology in 7 clusters. This classification had the higher rate of allocation of farms in one cluster and was in accordance with the main characteristics defined by experts. The classification of French pig farms provided in table 1, gives insights on the diversity of pig rearing activity in France. The French pig production is supplied by farms specialized in pig, crops, cattle or dairy production. The scale of these farms ranges from small to large. Thus, it is of

[5] Amounts of nitrogen deposited on land surfaces from the atmosphere are not included due to a lack of data regarding this variable.

relevance to take into account these structural differences for an efficiency analysis of the pig sector as a whole.

In order to check the validity and stability of this classification, several statistical analyses were implemented. Results are provided for year 1996.

The first statistical analysis was a *Kruskal–Wallis test*, which is a generalization of a Wilcoxon test in the case of several samples. It tests as the null-hypothesis the assumption that the distribution of a variable is identical among each cluster and as the opposite hypothesis, the assumption that the distribution of one variable is different at least in one cluster. The null-hypothesis was rejected for all variables used. The second test was a *Wilcoxon test*, which considers the same null-hypothesis and tests it two by two classes. These tests also rejected the null-hypothesis. Between two classes, there always have one variable that has a different distribution from the others.

A *discriminant analysis* was implemented to test for the misclassification of farms within the 7 clusters. Three methods were used. The *resubstitution method* provides an estimate of misclassification rates using the original dataset to define the classification rules and to test them. The *leaving one out method* gives a realistic estimate of misclassification rates by using a sample without one observation to define the classification rules and testing them on that observation. The *cross-validation method* provides an estimate of true misclassification rates. The classification rules were defined with the help of a sub-sample accounting for 25% of the total sample and were tested on the other sub-sample. Results are presented in the table 2.

Table 1. Classification of French farms with pig farming activity in 7 groups

Class 1	Large-scale farms specialized in crop production
Class 2	Medium-scale farms specialized in crop production
Class 3	Large-scale farms specialized in pig production
Class 4	Medium-scale farms specialized in pig production
Class 5	Large-scale farms specialized in dairy production
Class 6	Small-scale farms specialized in cattle and dairy production
Class 7	Small-scale farms specialized in cattle production

Table 2. Misclassification rates (%) in 1996

Class 1	Class 2	Class 3	Class 4	Class 5	Class 6	Class 7
Resubstitution method						
7.50	19.54	6.45	20.93	10.91	14.69	4.62
Leaving one out method						
7.50	22.99	16.13	23.26	20.00	17.51	9.23
Cross-validation method						
11.11	26.92	10.00	41.67	22.22	14.29	38.46

Misclassification rates are less than 20 % for the clusters 1, 3 and 6, while they are upper than 20 % as for classes 2, 4 and 5. For class 7, only the cross-validation method provides a high rate of misallocation. Thus, some classes appear to be very specific as classes 1, 3, 6 or 7. It seems more difficult to allocate farms into classes 2 and 4. It may be expected a superposition of the limits of these two clusters. This could be a direction for a potential structural evolution in the pig sector.

DEA Results

Individual farm's economic and environmental efficiency measurement

All previous statistical analysis aim at allocating farm within a homogenous technological group that characterizes a homogenous productive structure. A DEA analysis can then be developed for each group. The average values for the productive and environmental efficiency measurements which are based on the directional distance functions as specified in (4) are reported in Table 3 for the year 1996. Results for years 1997 to 2000 are provided in Appendix A2.

On average, the productive and environmental efficiency of farms is 1.196 in 1996. Thus, desirable outputs (farm's revenues) can be increased by 19% jointly with a decrease in the production of undesirable outputs (nitrogen emissions) by the same amount. However, as shown in Table 3 this result hides a high variability among farms since some of them are efficient with a score of unity while some others are inefficient with a maximum inefficiency score of 2.572. In the former case, it means that the farm can continue to behave as observed since its production bundle is not over performed by another farm or a linear combination of other farms. This means that based on current observed best-practices it is impossible to suggest a direction for an efficiency improvement for these farms. In the latter case, the value indicates that the most inefficient farm can increase its revenue by 157% and decrease its nitrogen level by the same amount to become efficiency. It can be done by implementing practices from observed efficient farms.

When considering the average productive and environmental efficiency of farms in each cluster for year 1996, clusters with the lower variability are those where farms are specialised in pig production and those where farms have an activity on a large scale and that are specialised in crop production (class 1) or in dairy production (class 5). Similar results are provided for years 1997 to 2000.

Thus, farms with a high degree of specialisation in pig production and a large scale for their activity have a lesser variability in their efficiency than farms with a small scale and a more diversified activity. Few improvements can be expected from the formers without of change in the production technology. However, high potential increases in revenues and decrease in nitrogen surplus can be reached by the other farms just by developing more efficient practices without any change in the current production technology.

Cluster and Sector economic and environmental efficiency measurement

The cluster efficiency measurement indicates the percentage increase in production and decrease in pollution that each group of farms could achieve if all farms within the cluster become efficient. It is not an average value as in table 3 but a value resulting from an aggregation process defined in (7). Thus, they provide a measurement of the efficiency of each farming system in presence of individual farms heterogeneity (table 4).

Table 3. Mean of individual efficiency measurement of farms within their own cluster

1996	*Mean*	*Std-dev*	*Min*	*Max*
1: Large-scale and crop production	1.095	0.113	1.000	1.347
2: Medium-scale and crop production	1.195	0.208	1.000	1.963
3: Large-scale and pig specialized	1.083	0.093	1.000	1.292
4: Medium-scale and pig specialized	1.057	0.084	1.000	1.360
5: Large-scale and dairy specialized	1.101	0.117	1.000	1.494
6: Small-scale and cattle and dairy specialized	1.276	0.241	1.000	2.572
7: Small-scale and cattle specialized	1.284	0.268	1.000	2.037
Mean	*1.196*	*0.219*	*1.000*	*2.572*

Results are variable over the time period and the pig farming systems. On average, clusters that are the most efficient are clusters 4 and 3 that are those specialized in pig production. They have an average efficient score of 7% and 9%, respectively. There have very few possibilities for improvement within these two classes in a direction that increase revenue and decrease nitrogen surplus based on the current technology they used. The higher efficiency score was in 1999 and the lesser one in 1998 for class 3 (large scale farms specialized in pig production). For the class 4 with a medium-scale for activity, years 1996 and 1999 are the most efficient. However, this group has a high level of efficiency with a potential improvement in revenue and reduction in nitrogen surplus that is lesser than 10% each year. It is the most efficient farming system.

Farming systems operating on a large scale and specialized either in crop production or dairy production have an average score of 13% and 14%, respectively. However, these figures hide different evolutions over years. For the group of farms specialised in crop production, best efficient scores are in 1996 and 2000 with values under 10% while worst efficiency scores are provided in 1997 and 1996 with values higher than 16%. Considering farms specialized in dairy production, the lower efficient score is valued at 8% in 1996 while the worst efficiency score is at 16% in 1998. For these farming systems, there are rooms for increase in revenues and decrease in nitrogen surplus. The highest potential improvements in efficiency can be expected by farming systems with a small-scale and diversified activity, mainly classes 6 and 7. Regarding class 7, the efficiency score is always higher than 20% with a maximum of 31% in 1998 and a minimum of 21% in 1999.

Thus the scale of activity and the degree of diversification appear as two main directions for improvement of economic and environmental efficiency of French pig farming system.

The sector efficiency measurement, provided in table 4, indicates the percentage increase in revenues and decrease in pollution that the pig sector as a whole could achieve if all farming systems become efficient by the aggregation process defined in (8). Thus, they provide a measurement of the efficiency at the sector level in presence of heterogeneous farming systems.

On average over years, the efficiency score is worth at 14%. The most efficient years are 1998 and 2000 with values between 12% and 14%. The most inefficient year is 1998 with a value at 17%. Results at the sector level show an increase in efficiency from 1996 to 1998 and then a decrease until 2000. This evolution is closely related to the economical context regarding this time period. Indeed, years 1996 to 2000 correspond to the highest part of the pig production cycle. From 1996 to 1998, producers were affected by a very important decrease in production prices. Prices were so low that production costs were not covered. Furthermore, a more restrictive environmental regulation was implemented in France in 1998. It affects the most polluted areas from France, with the most intensive and concentrated pig production systems, mainly the western part of France.

Results at the sector level illustrate some important points. The efficiency measurements reflect the impact of the economical context and the strength of the regulation on manure management. This evolution had implied a technical improvement of farms' efficiency. The additional cost involved by production price decrease and by manure management regulation was best integrated by farming systems operating at the largest-scale of production than at the smallest-scale ones. However, farming systems with cattle activities or with a small and medium-scale have the highest potential for improvements both in economical and environmental dimensions since they have the lowest efficiency scores.

Table 4. Cluster and Sector efficiency measurement

Cluster efficiency measurement	*1996*	*1997*	*1998*	*1999*	*2000*	*Mean*
1: Large-scale and crop production	1.092	1.167	1.148	1.163	1.072	*1.128*
2: Medium-scale and crop production	1.150	1.136	1.140	1.228	1.141	*1.159*
3: Large-scale and pig specialized	1.066	1.109	1.124	1.059	1.090	*1.089*
4: Medium-scale and pig specialized	1.045	1.098	1.048	1.081	1.089	*1.072*
5: Large-scale and dairy specialized	1.088	1.142	1.166	1.159	1.149	*1.141*
6: Small-scale and cattle and dairy specialized	1.229	1.253	1.331	1.210	1.227	*1.250*
7: Small-scale and cattle specialized	1.187	1.138	1.213	1.156	1.119	*1.163*
Sector efficiency measurement	*1.122*	*1.146*	*1.173*	*1.154*	*1.128*	*1.143*

CONCLUSION

This paper is an attempt to provide an evaluation of the economic and environmental efficiency of the French pig sector by taking into account the heterogeneity of farming systems producing pigs in France and the degree of efficiency of farms in the implementation of their production technology. The measurement is built on a three-stage approach.

Firstly, a cluster analysis was implemented to provide a classification of main farming systems producing pigs in France. It provided seven clusters of farms with a homogenous productive structure, built on three items: land use, labour use and diversity of production versus degree of specialization in pig production. The stability of the classification was studied from 1996 to 2000 by using statistical techniques. Secondly, a nonparametric frontier analysis, called Data Envelopment Analysis (DEA), which allows for efficiency measurement was used. The efficiency evaluation is implemented in both economic and environmental terms. Thirdly, an aggregation process provides an evaluation of the efficiency at the farming system and the sector level.

Main results show an average sector inefficiency of 14% with a range from 12 to 17%. Farming systems that are the most efficient are those specialized in pig production whatever the scale of their activity. They have very few possibilities for an improvement with the current observed technology they implement, even if they do it with a high degree of efficiency and a low degree of heterogeneity between farms. The only way for these farms to improve their revenues jointly with a reduction in their nitrogen surplus would be a change from their current technology to a more environmentally-friendly and still competitive one. Without any change in technology, improvements can be expected from farming systems which are diversified and that operate at a smaller scale. They have rooms for increasing their revenues and contributing to the reduction of nitrogen surplus.

Thus, there are possibilities for improvements in economic and environmental efficiency in the pig sector without change in technology. However, all the farming systems are not concerned. It is just for those that are diversified with a small or medium scale of production. For the others, a change in their production technology is necessary for increasing their revenues together with a decrease in their polluting impact. These results are of interest for the design of regulations in this sector.

REFERENCES

Backus, G., van Wagenberg, C. & Verdoes, N. (1998). Environmental Impact of Pig Meat Production, *Meat Science*, *49*, *1*, 65-72.

Ball, V. E., Lovell, C. A. K., Nehring, R. F. & Somwaru, A. (1994). Incorporating undesirable output into models of production: An application to the US agriculture, *Cahiers d'Economie et Sociologie Rurales*, *31*, 60-74.

Charnes, A., Cooper, W. W. & Rhodes, E. (1978). Measuring the Efficiency of Decision Makimg Units, *European Journal of Operational Research*, *2*, 429-444.

Cooper, W. W., Seiford, L. M. & Zhu, J. (2004). *Handbook on Data Envelopment Analysis.* Springer.

Chung, Y. H., Färe, R. & Grosskopf, S. (1997). Productivity and Undesirable Outputs: A Directional Distance Function Approach, *Journal of Environmental Management*, *51*, *3*, 229-240.

Emrouzenejad, A., Parker, B. R. & Tavares, G. (2008), Evaluation of research in efficiency and productivity: A survey and analysis of the first 30 years of scholarly literature in DEA, *Socio-Economic Planning Sciences*, *42*, *3*, 151-157.

Färe, R. & Grosskopf, S. (2004). *New Directions: Efficiency and Productivity.* Kluwer Academic Publishers.

Färe, R., Grosskopf, S. & Lovell, C. A. K. (1994). *Production frontiers.* Cambridge: Cambridge University Press.

Färe, R., Grosskopf, S. & Tyteca, D. (1996). An activity analysis model of the environmental performance of firms – application to fossil-fuel-fires electric utilities, *Ecological Economics*, *18*, 161-175.

Farrell, M. J. (1957). The Measurement of Productive Efficiency, *Journal of the Royal Statistical Society*, *120*, 449-460.

Førsund, F. R. & Sarafoglou, N. (2002). On the Origins of Data Envelopment Analysis, *Journal of Productivity Analysis*, *17*, 23-40.

Gattoufi, S., Oral, M. & Reisman, A. (2004). Data Envelopment Analysis Literature: A Bibliography Update (1951-2001), *Socio-Economic Planning Sciences*, *38*, 159-229.

Norris, P. & Thurow, A. (1999). Environmental Policy and Technology Adoption in Animal Agriculture. In: F., Casey, A., Schmitz, S. Swinton, & D. Zilberman, (Eds.), *Flexible Incentives for the Adoption of Environmental Technologies in Agriculture.* Kluwer Publishing.

OECD, (2003). *Agriculture, Trade and the Environment, The Pig Sector.* Paris: OCDE.

Seiford, L. M. & Thrall, R. (1990). Recent Developments in DEA: The Mathematical Programming Approach to Frontier Analysis, *Journal of Econometrics*, *46*, 7-38.

Seiford, L. M. (1997). A Bibliography for Data Envelopment Analysis (1978-1996), *Annals of Operations Research*, *73*, *19*, 393-438.

Tyteca, D. (1997). Linear programming models for the measurement of environmental performance of firms – concepts and empirical results, *Journal of Productivity Analysis*, *8*, 183-198.

APPENDIX A1

We assume that there are $j=1,\ldots,J^k$ observations of inputs and outputs (x_j, y_j, b_j) in the cluster k $(k=1,\ldots, K)$. Following Färe *et al.*, 1994, we can construct the following output set:

$$P^k(x) = \Big\{(y,b) : \sum_{j=1}^{J^k} z_j y_{jm} \geq y_m \quad m=1,\ldots,M$$

$$\sum_{j=1}^{J^k} z_j b_{js} = b_s \quad s=1,\ldots,S$$

$$\sum_{j=1}^{J^k} z_j x_{jn} = x_n \quad n=1,\ldots,N$$

$$\sum_{j=1}^{J^k} z_j = 1$$

$$z_j \geq 0 \quad j=1,\ldots,J^k \Big\}$$

In this formulation, the inequalities for inputs and desirable outputs make them freely disposable. The undesirable outputs are modelled with equalities. This makes them weakly disposable. Furthermore, the intensity variables

z_j, $j = 1,\ldots,J^k$ are taken to be nonnegative. This implies that the production technology exhibits constant returns to scale (CRS).

For each observation j within the cluster k, the output directional distance function is computed as the solution to a linear programming problem[6]. For example, for j',

$$\vec{D}_o^k(x_{j'}, y_{j'}, b_{j'}; y_{j'}, -b_{j'}) = \max \beta_{j'}^k$$

$$\begin{aligned}
s.t. \quad & \sum_{j=1}^{J^k} z_j y_{jm} \geq (1+\beta_{j'}^k) y_{j'm} && m = 1,\ldots,M \\
& \sum_{j=1}^{J^k} z_j b_{js} = (1-\beta_{j'}^k) b_{j's} && s = 1,\ldots,S \\
& \sum_{j=1}^{J^k} z_j x_{jn} \geq x_{j'n} && n = 1,\ldots,n \\
& \sum_{j=1}^{J^k} z_j = 1 \\
& z_j \geq 0 \quad j = 1,\ldots,J^k
\end{aligned}$$

APPENDIX A2

Mean of individual efficiency measurement of farms within their own cluster

1997	*Mean*	*Std-dev*	*Min*	*Max*
1: Large-scale and crop production	1.182	0.187	1.000	1.784
2: Medium-scale and crop production	1.209	0.233	1.000	1.898
3: Large-scale and pig specialized	1.119	0.105	1.000	1.394
4: Medium-scale and pig specialized	1.113	0.115	1.000	1.345
5: Large-scale and dairy specialized	1.181	0.159	1.000	1.621
6: Small-scale and cattle and dairy specialized	1.300	0.276	1.000	2.576
7: Small-scale and cattle specialized	1.205	0.210	1.000	1.995
Mean	*1.207*	*0.217*	*1.000*	*2.576*

[6] For solving linear programs, the procedure lp from SAS/OR (Operational Research) was used.

1998				
1: Large-scale and crop production	1.188	0.215	1.000	1.811
2: Medium-scale and crop production	1.225	0.294	1.000	2.597
3: Large-scale and pig specialized	1.132	0.129	1.000	1.517
4: Medium-scale and pig specialized	1.054	0.098	1.000	1.368
5: Large-scale and dairy specialized	1.194	0.171	1.000	2.130
6: Small-scale and cattle and dairy specialized	1.399	0.301	1.000	2.365
7: Small-scale and cattle specialized	1.261	0.208	1.000	1.860
Mean	*1.246*	*0.252*	*1.000*	*2.597*
1999				
1: Large-scale and crop production	1.247	0.499	1.000	4.496
2: Medium-scale and crop production	1.290	0.290	1.000	2.122
3: Large-scale and pig specialized	1.072	0.098	1.000	1.323
4: Medium-scale and pig specialized	1.084	0.103	1.000	1.299
5: Large-scale and dairy specialized	1.174	0.125	1.000	1.554
6: Small-scale and cattle and dairy specialized	1.249	0.225	1.000	1.979
7: Small-scale and cattle specialized	1.217	0.212	1.000	1.976
Mean	*1.213*	*0.261*	*1.000*	*4.496*
2000				
1: Large-scale and crop production	1.080	0.105	1.000	1.352
2: Medium-scale and crop production	1.187	0.199	1.000	1.723
3: Large-scale and pig specialized	1.126	0.134	1.000	1.454
4: Medium-scale and pig specialized	1.090	0.099	1.000	1.416
5: Large-scale and dairy specialized	1.165	0.129	1.000	1.549
6: Small-scale and cattle and dairy specialized	1.267	0.215	1.000	1.868
7: Small-scale and cattle specialized	1.200	0.329	1.000	3.364
Mean	*1.177*	*0.201*	*1.000*	*3.364*

In: Manure: Management, Uses and… ISBN: 978-1-61668-424-2
Editors: C. S. Dellaguardia, pp. 145-170

Chapter 5

Similar Bacterial Community Structure and High Abundance of Sulfonamide Resistance Genes in Field-Scale Manures

***Chu Thi Thanh Binh*[1], *Holger Heuer*[1], *Newton Carlos Marcial Gomes*[2], *Martin Kaupenjohann*[3] *and Kornelia Smalla*[1*]**

[1]Julius Kühn-Institut, Federal Research Centre for Cultivated Plants (JKI), Messeweg 11-12, D-38104 Braunschweig, Germany
[2]CESAM and Department of Biology, Universidade de Aveiro, Campus Universitário de Santiago, 3810-193 Aveiro, Portugal
[3]Institute of Ecology, Berlin University of Technology, Salzufer 11-12, D-10587 Berlin, Germany

Abstract

The bacterial community structure and antibiotic resistance of bacteria in 16 spread manures originating from 15 pig producing facilities in Germany were investigated. The farms varied in numbers of pigs,

* Corresponding author. Email: kornelia.smalla@jki.bund.de, Institut für Epidemiologie und Pathogendiagnostik, JKI; Messeweg 11-12, 38104 Braunschweig, Germany; Tel: +49-531-2993814; Fax: +49-531-2993006.

production type, and antibiotic usage. Between 2 to 69% of cultured bacteria were resistant to the recalcitrant sulfonamide antibiotic sulfadiazine, while less than 0.3% showed resistance to the much less stable β-lactam antibiotic amoxicillin. The abundance of the β-lactam resistance gene *bla*-TEM, and the sulfonamide resistance genes *sul1* and *sul2* in total community DNA was quantified by real time PCR relative to 16S rRNA gene copies. In all manures, a high proportion of bacteria carried *sul* genes, especially *sul2*, while the relative abundance of *bla*-TEM was at least two log units lower.

Bacterial community fingerprinting by PCR-DGGE of 16S rRNA gene fragments revealed that the dominant bacterial populations were surprisingly similar in all the manure samples. In contrast, enterobacterial patterns varied much more between samples, indicating that this group of potential pathogens is more susceptible to factors differing among piggery manures. Cloning and sequencing of 16S rRNA gene fragments reamplified from dominant DGGE bands, and of 16S rRNA genes amplified from total community DNA showed that the dominant populations in manure belonged to the Clostridia. In conclusion, substantial amounts of antibiotic resistant bacteria are released into the environment by spread manure, which contributes to the cycling of resistance genes between animal production facilities, human communities and clinical environments, resulting in an increasing threat to antibiotic therapy.

INTRODUCTION

Manure fertilization is one of the measures in agriculture to maintain soil fertility. However, the use of veterinary antibiotics in animal husbandry is suspected to affect the bacterial community structure in manure and thereby increase the prevalence of antibiotic resistance of the bacteria in manure [66]. Recently, more and more studies showed that manure seems to be a pool for antibiotic resistant bacteria [18, 26, 44, 57], of antibiotic resistance genes [7, 34, 36, 67] and of transferable plasmids carrying antibiotic resistance genes [32, 49, 59].

Among the most frequently used antibiotics for animal husbandry are sulfonamide and β-lactam antibiotics that are used to control diarrhea and various other infectious diseases [43, 56]. Resistance towards these antibiotics can emerge in the gut microflora of treated animals [40]. Depending on the chemical structure of the antibiotic, a substantial part can be excreted and thus enter the soil via manure [27, 30, 39]. Three genes (*sul1*, *sul2*, *sul3*) were found to confer sulfonamide resistance by encoding dihydropteroate synthases which are sulfonamide-insensitive [51, 53, 62]. Many genes have been found to encode proteins involved in β-lactam antibiotic resistance [11, 47] but *bla*-TEM genes are the most abundant β-lactamase encoding genes [41]. *Sul* and *bla*-TEM genes were mostly investigated in clinical isolates [4, 14, 20, 25]. Furthermore, *bla*-TEM genes were also detected in bacterial isolates or total community DNA (TC-DNA) from aquatic environments [28, 29], and *sul* genes were detected from environmental *Salmonella* isolates [3]. Only in a few studies, quantification of these genes in environmental flora was attempted [7, 34, 50]. However, no quantitative data for *sul* and *bla*-TEM genes in piggery manure used for soil fertilization were available.

The effect of antibiotics introduced via manure into soils on soil microbial communities is presently investigated in the frame of a DFG research group [8, 34, 38]. In order to provide baseline data on bacterial communities typical for manure used for soil fertilization which could also serve as a reference, a survey of a unique set of 16 manures collected from 15 different pig producing facilities (PPF) in Germany (so-called field-scale manure) was performed to achieve insights into the abundance of antibiotic resistance genes, mobile genetic elements and the bacterial community structure in these manures. In this study we quantified the abundance of *sul1, sul2* and *bla*-TEM genes by real time PCR. Furthermore, the structural diversity of the 16 manure samples based on 16S rRNA genes amplified from TC-DNA was analyzed by denaturing gradient gelelectrophoresis (DGGE) and sequence analysis. The data of this study provide for the first time information on the bacterial community structure and the abundance of *sul* and *bla*-TEM resistance genes in a unique set of field-scale manures collected from PPF that represented diverse sizes of herds, meat or piglet production, feeding and antibiotic usage.

Materials and Methods

Manure and Sampling

Sixteen manure samples were taken from 15 different PPF in Germany in May and June 2006. In each PPF, the manure was collected into storage tanks for approximately 6 months. Before the manure was spread onto fields, it was intensively stirred and samples were taken. Characteristics of PPF and chemical composition of manures were previously described [9].

Cultivation-Dependent Analysis

One gram of manure was resuspended in 9 ml sterile saline, vortexed and serial dilutions were spread onto R2A medium (Merck) supplemented only with cycloheximide (100 mg l^{-1}) for total CFU (colony forming units) counts, or additionally with one of the two antibiotics AMX (100 mg l^{-1}) or SDZ (100 mg l^{-1}) for counts of AMX or SDZ resistant bacteria, respectively. Plate counts were determined after incubation at 28°C for 3 days and expressed as resistance quotient RQ (%) = CFU g^{-1} of antibiotic resistant bacteria / CFU g^{-1} of total bacteria x 100.

Quantification of *bla*-TEM and *sul* Resistance Genes

TC-DNA was extracted from 1 g manure by using the FastDNA® SPIN Kit for soil, and purified by the GENE CLEAN® SPIN Kit according to the manufacturer's protocols (MP Biomedicals, Illkirch, France). Primers and probe targeting *bla*-TEM for quantification of this gene by a 5'nuclease assay were designed using the program Primerexpress (Applied Biosystems, Darmstadt, Germany). The primers amplify a 56-bp fragment of the TEM gene of pUC19. The specificity of primers and probe was checked against 200 sequences of *bla*-TEM types in GenBank. The *bla*-TEM, *sul1*, *sul2* and bacterial 16S rRNA genes from 16 manures were quantified by real-time PCR using the ABI PRISM 7000 sequence detection system (Applied Biosystems). Primers for *bla*-TEM, *sul2* and the TaqMan probe labelled 5' with FAM and quenched 3' by TAMRA are shown in Table 1. Each PCR mixture (50 µl) consisted of 1 X buffer II, 1.25 U of AmpliTaq Gold (Applied Biosystems), 2.5 mM $MgCl_2$ (3.75 mM for *sul2*), 0.2 mM dNTPs, 0.25 µM of each primer and the TaqMan probe. The temperature profiles were 95°C for 10 min and 40 cycles of 95°C 15 s, 60°C 1 min for *bla*-TEM, or 40 cycles of 94°C 15 s, 51°C 15 s, 60°C 1 min for *sul2*. The primers and probe used for 16S rRNA genes quantification were described by Takai and Horikoshi [63]. Conditions for qPCR of 16S rRNA and *sul1* genes were described by Heuer and Smalla [34]. Serial dilutions of pUC19 vector ($1.7x10^{11}$ molecules) and gel-purified PCR product from R388 *sul1* (995bp, OD_{260}=0.538), RSF1010 *sul2* (965 bp, OD_{260}=0.175), and *E. coli* 16S rRNA genes (1506 bp, OD_{260}=0.533) were used as templates for standard curves. The log_{10} of *bla*-TEM, *sul1*, and *sul2* target genes were calculated relative to 16S rRNA genes.

Table 1. Primers and probes

Target	Oligo	Sequence (5' to 3')	T_a	Study
bla-TEM	TEM175F	CGCCCCGAAGAACGTTTT	60°C	This study
	TEM231R	CGCGCCACATAGCAGAACTT		
	tpTEM	CAATGATGAGCACTTTT		
sul1	qSUL653F	CCGTTGGCCTTCCTGTAAAG	60°C	[34]
	qSUL719R	TTGCCGATCGCGTGAAGT		
	tpSUL1	CAGCGAGCCTTGCGGCGG		
sul2	qsul2-595F	CGGCTGCGCTTCGATTT	53°C	[30]
	qsul2-654R	CGCGCGCAGAAAGGATT		
	tpsul2-614	CGGTGCTTCTGTCTGTTTCGCGC		
Bacteria 16S rRNA gene	Bact348f	AGGCAGCAGTDRGGAAT	50°C	[63]
	Bact786r	GGACTACYVGGGTATCTAAT		
	tpBact	TGCCAGCAGCCGCGGTAATACR DAG		
Enterobacteria 16S rRNA gene	Entero-F234	GATGWRCCCRKATGGGA	57°C	This study
	Entero-R1423	AKCTAMCTRCTTCTTTTGCAA		

Bacterial and Enterobacterial 16S rRNA Gene PCR-DGGE

Bacterial 16S rRNA gene fragments (primers: F984GC, R1378) were amplified from TC-DNA as described by Heuer *et al.* [33]. Polyacrylamide gels for DGGE had a denaturing gradient of 26 to 58% urea/formamide. Polymerization was initiated with 0.17% (v/v) TEMED and 0.05% (w/v) ammonium persulfate. Gels were run for 6 h, 220 V in 0.5 x TAE at 58°C in a DCode system (Bio-Rad, München, Germany). DNA was visualized by silver staining [35].

A nested-PCR was applied to amplify 16S rRNA gene fragments specific for the family *Enterobacteriaceae*. Degenerate primers for amplifying 16S rRNA genes of *Enterobacteriaceae* (F234/R1423) were designed using the PROBE DESIGN and MATCH PROBE subroutines in the ARB software [42] (Table 1). The Probe Match function of the Ribosomal Database Project II (http://rdp.cme.msu.edu/) was used for *in silico* analysis of the primer specificity. The searches were done based on the last ten 3' -end nucleotides of each primer (region with the greatest template discrimination). The primers were empirically tested with genomic DNA of *E. coli*. The reaction mixture for PCR amplification (25 μl) was prepared containing 1X TrueStart PCR buffer, 2.5 U TrueStart polymerase (Fermentas, St. Leon-Rot, Germany), 0.2 mM dNTPs, 3.75 mM $MgCl_2$, 2.5 μg BSA, 5% (v/v) DMSO, 0.4 μM of each primer. The temperature profile was as follows: initial denaturation (94°C for 5 min), 30 cycles of denaturation (94°C for 1 min), annealing (57°C for 1 min), extension (72°C for 2 min), and final extension (72°C for 7 min). Afterwards, the PCR products were diluted ten times in water and 1 µl was used for PCR amplification of 16S rRNA fragments (F984GC/R1378) for DGGE as described above. DGGE gels were transmissively scanned with high-resolution settings (Epson 1680 Pro, Seiko-Epson Corp. Suwa, Nagano, Japan). The community fingerprints were analyzed using GelCompar 4.0 (Applied Maths, Ghent, Belgium). The Pearson correlation index (r) for each pair of lanes within the gel was calculated as a measure of similarity between the community fingerprints. Cluster analysis was performed by applying the unweighted pair group method using average linkages (UPGMA) to the matrix of similarities.

Cloning and Sequence Analysis of 16S rRNA Genes

Dominant bands, which occurred commonly among DGGE patterns of 16S rRNA gene fragments of bacteria in general and of *Enterobacteriaceae* in particular amplified from manures, were excised from gels (Figure 3 and Figure 4), reamplified, cloned and sequenced. The pieces excised from the gel were placed in 20 µL MilliQ water, thoroughly ground with a pipette tip and stored 24 h at 4°C to allow the DNA to diffuse out of the gel. Two µL of eluted DNA were used as the DNA template for 16S rRNA gene amplification according to Heuer *et al.* [33] and checked by DGGE to confirm that the amplicon had the same electrophoretic mobility as the excised band. Two µL of DNA eluted from excised bands were used for 16S rRNA gene amplification (without GC clamp), purified and then cloned into the pGEM-T vector (Promega, Madison, WI, USA) and transformed into *E. coli* JM109. Approximately 25 white colonies were picked per cloned band, resuspended in 50 µl MilliQ water, boiled and 1 µL was used as template for 16S rRNA gene amplification as described by Heuer *et al.* [33]. Amplicons were run on a DGGE gel. The clones (one to three), which showed the same electrophoretic mobility as the band excised from the community patterns, were used as template for PCR amplification with primers SP6/T7 (Promega, Mannheim, Germany). PCR products were purified and sent for sequencing.

Cloning and sequencing was also done for 16S rRNA gene fragments directly amplified from TC-DNA of manures. While the PCR products obtained from TC-DNA of manures 6 or 12 were cloned separately, the amplicons from TC-DNA of the manures 1 and 4 were combined and those from amplicons of manures which were more than 76% similar according to clustering by GelCompar, were mixed together and then cloned into pGEM-T vector. One µl boiled cells of white colonies was used as template for PCR amplification with primers SP6/T7. PCR products from 24 clones of each group were purified and sequenced. The 16S rRNA gene fragments amplified from TC-DNA of manures 1 and 10, using the primers targeting *Enterobacteriaceae* were cloned and sequenced as described above.

The sequences were subjected to BLAST-N searches to find the most similar sequences in the GenBank database. The sequences, which could not be assigned to a particular genus based on the BLAST-N hit were classified at the genus level with the Naive Bayesian rRNA Classifier (Version 1.0) of the Ribosomal Database Project II (RDP-http://rdp.cme.msu.edu/) with a 95% limit of confidence. Sequences reported in this paper have the accession numbers from AB433995 to AB434095 in the DDBJ/EMBL/GenBank nucleotide sequence database.

RESULTS

Cultivation-Dependent Analysis of AMX and SDZ Resistance

In order to investigate the fraction of bacteria resistant to AMX and SDZ, serial dilutions of manure samples were plated onto R2A agar plates supplemented with cycloheximide (100 mg l^{-1}) and AMX (100 mg l^{-1}) or SDZ (100 mg l^{-1}). The resistance quotient given as percent of AMX or SDZ resistant counts to total counts of bacteria varied among manures, ranging from 2 to 69% while those of AMX were between 0.006 to 0.3%. Both AMX and SDZ resistance quotients were highest in manure 8 (Figure 1). A trend was observed that the average values of AMX and SDZ resistance quotients of manures from sows and piglets facilities (manures 1, 2, 3, 4, 8) were higher (0.97% and 41%, respectively) than those from meat production facilities (0.46% and 32%, respectively).

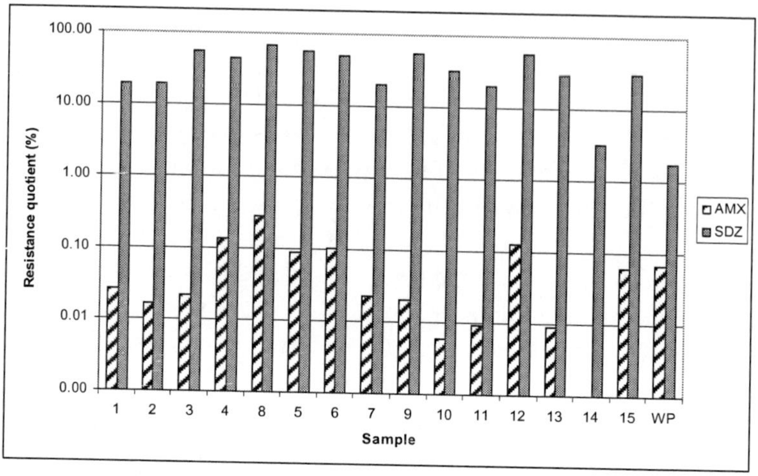

Figure 1. Resistance quotient of AMX and SDZ in 16 field-scale manures.

Cultivation-Independent Analysis of AMX and SDZ Resistance

Quantification of *bla*-TEM, *sul1* and *sul2* genes was done to evaluate the abundance of AMX and SDZ resistance genes in manures (Figure 2a). *Sul* genes were much more abundant than *bla*-TEM genes. Among the three antibiotic resistance genes quantified the *sul2* gene was most abundant with a range from log 5.5 (in manure 6) to log 9 (in manure WP) copies per gram manure. The abundance of *sul1* genes, ranged from log 4.4 (manure 6) to log 7.5 (manure 11 and WP). The *sul1* gene was less abundant than *sul2* in almost all manures (except for the manure 11) with an average value of log 6.43 molecules per gram manure in comparison with log 7.54 for *sul2*. In contrast to the *sul* genes the *bla*-TEM gene was much less abundant, with only log 2.64 (manure 1) to log 4.16 (manure WP) molecules per gram manure. The *bla*-TEM, *sul1* and *sul2* gene copy numbers were related to the bacterial 16S rRNA gene copies determined by qPCR (Figure 2b). In this way, the relative abundance of the resistance genes could be evaluated and compared among the manures. For *sul1*, the highest relative abundance was found in manure 9 with -1.2 log (copies *sul1*:16S rRNA gene) and least abundance in manure 13 with -4.5 of this value. For *sul2*, the highest and the lowest abundance was -0.3 and -3.2 for manure 1 and 13, respectively. Also for the *bla*-TEM gene, the lowest proportion was found for manure 13 and 14 (-7.3) and the highest (-4.5) for manure 1. The average values indicated a trend that the relative abundances of *sul1*, *sul2*, and *bla*-TEM in manures from sows and piglets were higher with -2.5, -1.3 and -5.2, respectively, than for manures from meat production facilities (-2.7, -1.7 and -5.5, respectively).

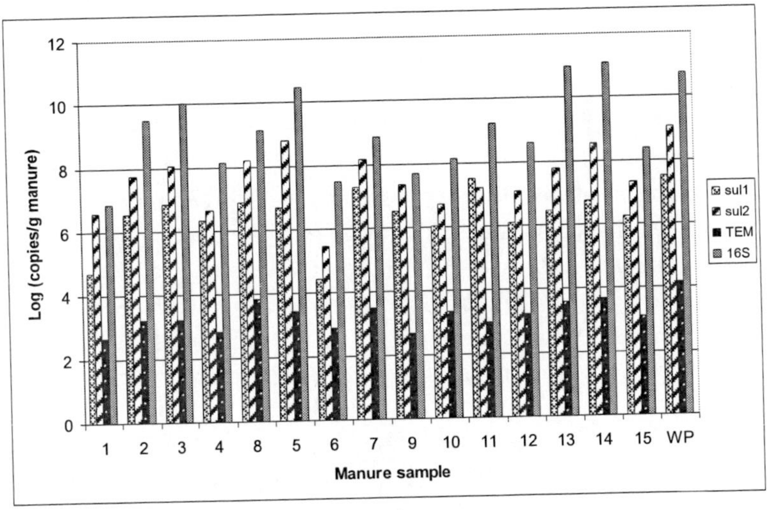

(a)

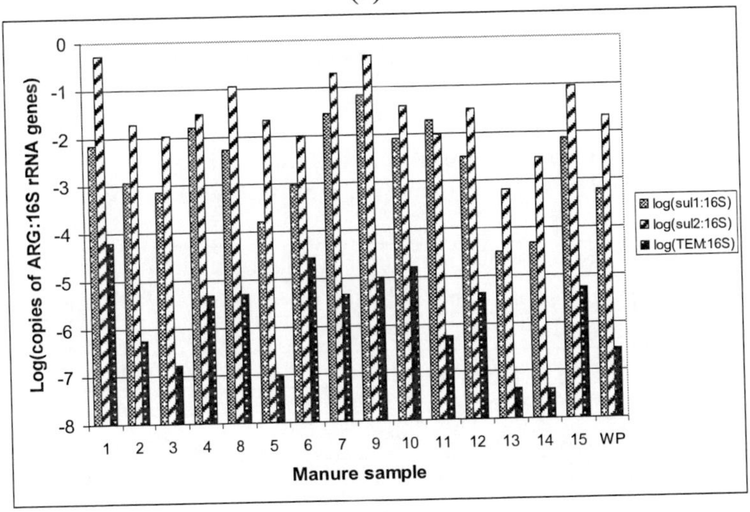

(b)

Figure 2. Quantification of *sul1, sul2* and *bla*-TEM genes in manure. ARG: antibiotic resistance genes. (a) Number of copies of these genes per g manure. (b) Abundance of these genes in manures relative to 16S rRNA gene copies.

Bacterial Community Structure Analysis

Bacterial 16S rRNA gene fragments were amplified from TC-DNA of manures and used for DGGE analysis. DGGE profiles of the manures were very similar (Figure 3). The cluster analysis showed that the DGGE profiles of 16S rRNA genes amplified from almost all manures were more than 75% similar (except for the manures 6 and 12) indicating that the manures contained similar dominant bacterial ribotypes. Some major bands appeared in DGGE profiles of almost all manures. Bands A, B, C and E were observed in the DGGE patterns of all manures. Although less predominant, band D also appeared in 9 among 16 manures. To investigate the dominant ribotypes in the manures, the major bands (A to E in Figure 3) were excised from DGGE gel, reamplified and cloned. Three clones comigrating with each band were sequenced. The results of sequencing, BLAST-N and RDP search are shown in Table 2. Sequences from band A had only 93% identity with *Acholeplasma* (belonging to phylum *Firmicutes,* class *Molicutes*, order *Acholeplasmatales*) according to BLAST-N search and then confirmed to affiliate with *Acholeplasma* according to RDP search. The sequences obtained for the cloned amplicons of bands B, C, and E showed the highest similarity to sequences of the genus *Clostridium*. Band D had 96% sequence identity to *Clostridium septicum*, and was assigned to unclassified *Clostridiaceae* by the RDP classifier.

Table 2. Sequences of common and dominant bands excised from DGGE gels of 16S rRNA gene fragments amplified from manures

Band	Closest organisms and/or environmental 16S rRNA gene	
	BLAST-N search (% identity)	RDP search
A	*Acholeplasma brassicae* - 93%	*Acholeplasma* sp.
B	*Clostridium disporicum* - 97%	*Clostridium* sp.
C	*Clostridium metallolevans* - 99%	*Clostridium* sp.
D	*Clostridium septicum* - 96%	unclassified *Clostridiaceae*
E	Uncultured *Clostridium* EU071510 - 99%	*Clostridium* sp.

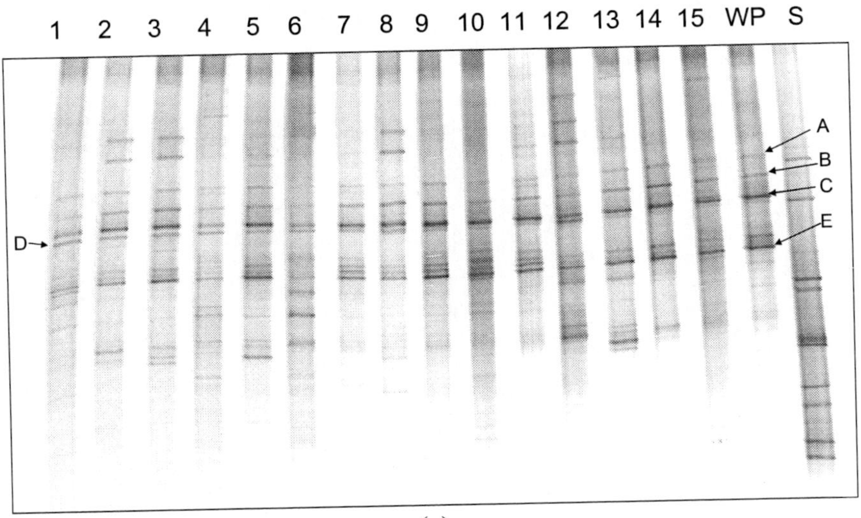

(a)

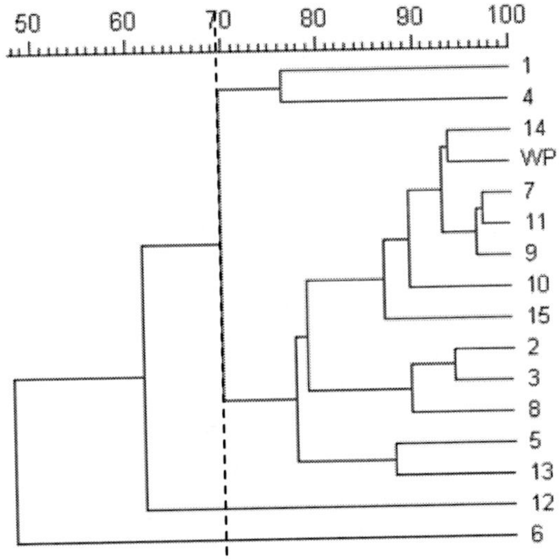

(b)

Figure 3. DGGE analysis of 16S rRNA gene fragments of 16 manures (a), and cluster analysis by UPGMA (b). S: standard. A, B, C, D and E indicate the bands which were excised, cloned and sequenced.

Table 3. Sequence analysis of cloned 16S rRNA genes amplified from TC-DNA of manures

Microorganism	Clone library				Total
	1+4	6	12	R	
Clostridia	19	14	10	13	***56***
Lactobacillus sp.	1		1	2	***4***
Streptococcus sp.		1	1		***2***
Turicibacter sp.		4		3	***7***
Unclassified *Firmicutes*		3	4	1	***8***
α- Proteobacteria	1				***1***
γ- Proteobacteria	1	2			***3***
Actinobacteria			1	1	***2***
Bacteroidetes				1	***1***
Verrucomicrobia				1	***1***
Unclassified bacteria	2		7	2	***11***
Total	***24***	***24***	***24***	***24***	***96***

In parallel, clone libraries were prepared from 16S rRNA gene amplicons for manures 6 and 12. In addition, amplicons from TC-DNA of manures 1 and 4 were combined to prepare library "1+4", and another library "R" represented all other manures which shared in the bacterial DGGE profiles more than 75% similarity (Figure 3). Twenty-four clones from each group were sequenced and searched by both BLAST-N and RDP for the most similar sequences. The sequencing results are summarized in Table 3. More than half of the sequences obtained belonged to the class *Clostridia* (56 of the 96 clones sequenced). Altogether, 77 clones were assigned to the phylum *Firmicutes* representing 80% of the total number of clones sequenced. The results of both the clone libraries and sequencing of dominant bands confirmed that *Firmicutes* in general, *Clostridia* in particular were the major bacterial inhabitants in all manures.

Enterobacterial Community Structure of Manures

In order to study the community composition of the hygienically relevant group of the *Enterobacteriaceae* in manure, primers specific for 16S rRNA genes of this group were designed and used for PCR amplification of these genes from TC-DNA of manures. The *in silico* analysis of the newly developed primers revealed that about 90% of 16S rRNA gene sequences from *Enterobacteriaceae* available in the RDP (11214 sequences covering the primer region) had a perfect match (last ten 3' -end nucleotide) with the forward *Enterobacteriaceae* primer Entero-F234. Less than 300 non-*Enterobacteriaceae* sequences showed a perfect match with this primer. The reverse primer (Entero-R1423) showed a perfect match with > 93% sequences from *Enterobacteriaceae* (5604 sequences covering the primer region). However, this primer was less specific, showing a perfect match with 1132 non-*Enterobacteriaceae* sequences. In general, our analysis indicated that the combination of both forward and reverse *Enterobacteriaceae* primers would exclude most of non-*Enterobacteriaceae* sequences and ensure specific amplification of this family from complex community DNA. The *in silico* analysis also indicated that among 48 genera belonging to *Enterobacteriaceae* family, five (*Arsenophonus*, *Buchnera*, *Phlomobacter*, *Plesiomonas* and *Thorsellia*) will not be targeted by the *Enterobacteriaceae* primer system developed in this work.

A subsequent PCR was done to amplify 16S rRNA gene fragments (F984GC, R1378) using PCR products of group-specific *Enterobacteriaceae* as template, then the PCR products were used for DGGE analysis. DGGE profiles of the bacteria belonging to the family *Enterobacteriaceae* in 16 manures were shown in Figure 4. In contrast to DGGE profiles of bacterial 16S rRNA gene fragments amplified from manures, which possessed very similar patterns, DGGE patterns of *Enterobacteriaceae* were more diverse.

In order to test the specificity of PCR amplification of this group, dominant bands from the DGGE gel were excised and cloned. BLAST-N search was done for all sequences to find the closest sequences in GenBank. The sequences that could not be classified to any genus after BLAST-N search were subjected to RDP search. Among 17 sequences retrieved from DGGE bands, 11 sequences showed 99-100% similarity to the bacteria belonging to *Enterobacteriaceae*. Among other 6 sequences, which were not determined as *Enterobacteriaceae,* three belonged to *Bacteroidetes*, one to *Clostridiaceae,* one to *Corynebacterium* and one to unclassified bacteria (Table 4).

In addition, nested PCR products of 16S rRNA gene fragments specific for *Enterobacteriaceae* from manure 1 and 10 were also cloned and sequenced. Among six clones sequenced (three sequences from each manure), five had 98-100% similarity to sequences of bacteria belonging to *Enterobacteriaceae* and one to *Clostridium* (Table 4).

Table 4. Sequence analysis of 16S rRNA genes amplified from total DNA of manure with primers designed to target *Enterobacteriaceae*

Band from manure	Number of clones sequenced	Closest phylogenetic relative according to	
		BLAST-N (Accession No.- % identity)	RDP classifier
1	1	*Kluyvera cryocrescens* (AM992189 - 99%)	
2	2		Unclassified *Enterobacteriaceae*
			Proteiniphilum (Bacteroidetes)
3	2	*Shigella sonnei* (EU723822 - 100%)	
			Unclassified *Bacteroidetes*
5	1		*Corynebacterium*
6	2	*Shigella sonnei* (EU723822 - 100%)	
8	2	*E. coli* (EU723821 - 100%)	
9	1	*Enterobacter*. sp (EU240200 - 99%)	
10	1		*Bacteroidetes*
12	1		Unclassified bacteria
14	2	*Pantoea ananatis* (EU331415 - 99%)	
			Unclassified *Clostridiaceae*
15	2	*Shigella sonnei* (EU723822 - 100%)	
M1	3	*Pectobacterium carotovorum* (EF530548 - 99%)	
M10	3	*Enterobacter* sp. (EU567059 - 98%)	
			Clostridium
		Shigella sonnei (EU723822 - 99%)	

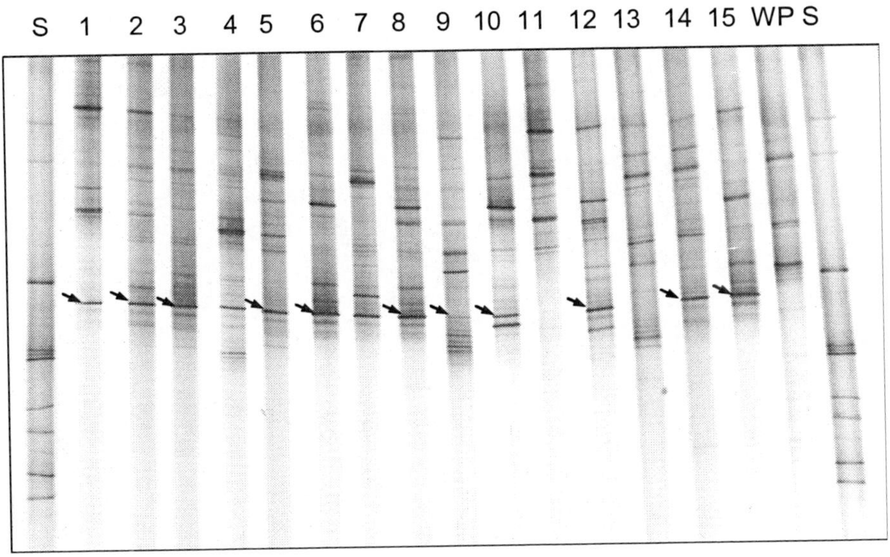

Figure 4. DGGE analysis of 16S rRNA gene fragments preferentially amplified from *Enterobacteriaceae* of 16 manures. S: standard. Arrows indicate the bands which were excised, cloned and sequenced.

DISCUSSION

The Abundance of Antibiotic Resistance in Manures

In all 16 manures studied, SDZ resistance among manure bacteria was much more prevalent than AMX resistance, and this corresponded to a much higher abundance of *sul* genes compared to *bla*-TEM genes. The long-term use of sulfonamides in animal husbandries and the frequent association of *sul* genes with mobile genetic elements [58] might have contributed to their high abundance in manure bacteria. Another important aspect might be the fate of sulfonamides used as veterinary medicines. In manure sampled from pigs treated with ^{14}C-SDZ over 10 days, more than 96% of the administered drug was found as parent compound or metabolites N-acetyl-SDZ and 4-hydroxy-SDZ. During storage of manure, the concentration of SDZ increased by 42% due to deacetylation of the metabolite N-acetyl-SDZ, while the minor metabolite 4-hydroxy-SDZ kept constant [30]. In this study, a higher proportion of *sul2* than of *sul1* genes was observed in TC-DNA for almost all manures. The abundance of *sul1* copies determined in TC-DNA of the 16 field-scale manures (on average log 6.43) was comparable to the number of *sul1* genes previously reported for manure which was stored for less than six months [34]. We did not attempt to quantify the third known *sul* gene (*sul3*) as this gene could not be detected by PCR and subsequent Southern blot hybridization in TC-DNA of all 16 manures indicating a low abundance or absence of this gene. Several cultivation-based studies on the abundance of *sul1* and *sul2* in Gram-negative isolates revealed frequencies of these genes within the same order of magnitude [4, 26, 54]. The frequency of the *sul* resistance genes found might depend on the taxonomy of the hosts and their localization on mobile genetic elements. The advantage of analyzing TC-DNA is that quantitative data on the abundance of *sul* or *bla*-TEM genes independent from their hosts and their cultivability can be provided. However, the identification of the host and localization on mobile genetic elements requires different experimental approaches such as cultivation or exogenous isolation, respectively. Previously Binh *et al.* [9] reported on the exogenous isolation of MGE from the same set of manures into rifampicin resistant *E. coli* recipients using selective media supplemented with AMX, SDZ or tetracycline. Replicon typing revealed that 28 of the 81 plasmids belonged to IncN, one to IncW, 13 to IncP-1ε, and 19 to the recently discovered LowGC-type plasmids [31]. IncP-1ε and the LowGC-type plasmids were also detected in TC-DNA of field-scale manures by PCR and Southern blot hybridization. The molecular analysis of MGE conferring SDZ resistance, which were exogenously captured into *E. coli* from the 16 field scale manures, revealed that among the 81 plasmids analyzed, 44 carried the *bla*-TEM, 23 the *sul1*, 46 the *sul2* and 6 the *sul3* gene [9]. Interestingly, in the study of Bibbal *et al.* [7] the abundance of *bla*-TEM genes in TC-DNA from manure was with 10^3-10^4 copies per gram manure much lower than the abundance of *bla*-TEM gene copy numbers determined in feces of ampicillin-treated pigs collected after one to seven days with ca. 10^6-10^9 copies per gram of feces. The *bla*-TEM genes are usually carried by Gram-negative bacteria [41] and the composition of bacteria has been shown to change remarkably during the storage of manure [52]. The considerably decreased abundance of enteric bacteria during storage of the manure [18] might be one reason why *bla*-TEM gene abundance in the manures used in our study was rather low. However, it seems that the incubation conditions during the filter mating activated bacteria carrying *bla*-TEM genes on MGE. The pig manure investigated in our study was collected for approximately 6 months in manure storage tanks and thus contained differently-aged manure. The low abundance of *bla*-TEM genes in the stored piggery manure might be not only the result of the changes in bacterial community during storage but is likely also the result of the different fate of β-lactam antibiotics administered to pigs in contrast to sulfonamides. Experimental data on the fate of amoxicillin in manure (M. Lamshöft, unpublished data) indicated a short half life most likely due to the omnipresence of β-lactamases. Recently, Binh *et al.* [10] amplified *aadA* genes localized on class 1 integrons from TC-DNA of all 16 field-scale manure samples. In addition, class 1 integrons without gene cassettes were detected in TC-DNA of many samples. Although not investigated in this study, the methods of manure storage and treatment may also have an impact on the persistence of antibiotic resistance genes [13, 67]. In those studies, the abundance of antibiotic resistance genes was substantially reduced in composted piggery manure. The quantity of tetracycline resistance genes in compost piggery manure determined by qPCR was 4.3 log per gram manure compared to 7.4 log in fresh manure observed by Yu *et al.* [67]. In the study by Chen *et al.* [13], the quantity of *erm* genes encoding resistance to macrolides-lincosamides-streptogramin B determined in compost piggery manure was lower than in fresh manure.

Another interesting observation was that in manure 13 originating from a PPF which did not use antibiotics, the abundance of all resistance genes quantified was lowest in comparison with that of other manures (Figure 2b). Interestingly, no plasmids could be exogenously isolated from manure 13 and also the abundance of broad-host-range plasmids detected in TC-DNA of manure 13 after PCR amplification was low [9], supporting the assumption that the abundance and transferability of antibiotic resistance genes is enhanced in response to antibiotic use. A trend for higher resistance quotients and abundances of antibiotic resistance genes was observed for piglet raising facilities rather than for meat production. This finding might be due to a more intense prophylactic usage of antibiotics in facilities for sows and piglets.

The sulfonamide resistance genes *sul1* and *sul2* were quantitatively detected in total DNA from river sediments by Pei *et al.* [50], using other qPCR systems. The study reported that the abundance of *sul* genes was much higher in sediments from sites with urban and agricultural impact than those from pristine sites demonstrating the relationship between human and agricultural activity and levels of antibiotic resistance genes in river sediments.

Similar Bacterial Community Structures in Diverse Manures with Clostridia Being a Major Component

The analysis of 16S rRNA genes amplified from TC-DNA of 16 manures by DGGE patterns and cloning and sequencing revealed that the dominant populations in manure after storage are *Clostridia* and that the population structure was quite similar and independent from the production type. The findings of our study are in agreement with the previously published data by Cotta *et al.* [15] which were based on bacterial isolates from piggery manure. Also this cultivation-based study showed that the dominant bacterial population in manure belonged to the *Firmicutes* including various *Clostridia.* By comparing the microbial population in piggery feces and stored manures using analysis of 16S rRNA genes, Whitehead and Cotta [65] showed that *Lactobacillus* and streptococcal species were most dominant in feces together with clostridial species, but after storage of the manure, the clostridial species became most abundant. The result of our study clearly showed that there is a typical structural composition of bacterial communities in manure as the 16S rRNA gene fingerprints of 16 manures were rather similar indicating similar dominant bacteria in all manure samples.

Among clones in the library, no sequences of bacteria belonging to *Enterobacteriaceae* were recovered, implying that these bacteria are minor populations in manure. However, from many previous studies, *Enterobacteriaceae* are known to be the hosts of a wide range of antibiotic resistance genes, especially *bla*-TEM and *sul* genes. Therefore, we attempted to have deeper insight into the diversity of this group. Interestingly, the DGGE patterns of this group were diverse among manure samples. Bacteria from different genera retrieved from bands of the same electrophoretic mobility also indicated the diversity of this group in manure. In agreement with the clone library, the non-*Enterobacteriaceae* sequences retrieved from dominant bands of the *Enterobacteriaceae* DGGE could be also an indication of low abundance of bacterial populations belonging to this family. As a consequence of the supposed low abundance only minute amounts of enterobacterial 16S rRNA gene fragments are amplified. Thus in the second PCR with GC-clamped primers also 16S rRNA gene fragments of dominant ribotypes not belonging to the Enterobacteriaceae are amplified resulting in unspecific amplification of 16S rRNA gene fragments as suggested before [23].

Reservoirs for *bla*-TEM and *sul* Genes in Manure?

Gram-negative bacteria are generally suspected to be the major reservoir for *bla*-TEM and *sul* genes [41, 58]. The cultivation-independent methods used in our study showed that Gram-positive bacteria belonging to *Clostridia* were dominant in all manures and that a high proportion of manure bacteria must have carried *sul* and *bla*-TEM genes. Although no attempt was made to isolate these dominant bacteria, which require anaerobic growth conditions, we assumed that those genes were carried also by *Clostridia*. Sulfonamides and AMX are among the commonly used antibiotics to treat enteritis caused by *Clostridium* in animal husbandry [43]. Although there are numerous reports on antibiotic resistance genes carried by *Clostridia* [2, 6, 16, 37, 61], or on the transfer of resistance genes from *Clostridium* to bacteria of different genera [45], *bla*-TEM genes have not yet been reported in *Clostridium*, and *sul* genes were very rarely found in *Clostridium*. Resistance of *Clostridia* to sulfonamides was already reported by Neikirk and Krieg [48], to amoxicillin by Bendle *et al.* [6].

CONCLUSION

Manure was identified as a source of pathogenic and antibiotic resistant bacteria [1]. These bacteria can be brought to soil when manure is used as fertilizer. Potentially pathogenic bacteria, e.g. *E. coli* and *Salmonella,* were found to persist in manured soil for some time [5, 22]. Manure contains considerable levels of antibiotic resistance genes carrying bacteria as shown in this study and by several others [13, 36, 67]. Often these antibiotic resistance genes are located on mobile genetic elements conferring antibiotic resistances [8, 59]. The 16 field-scale manures analyzed in this study were collected from 15 different pig producing facilities in Germany representing diverse sizes of herds, meat or piglet production. These manures were spread on fields, thereby introducing large amounts of bacteria carrying antibiotic resistance genes on mobile genetic elements into agricultural soils. The fate of the antibiotic resistance genes might depend on a variety of different abiotic and biotic factors. Bacteria carrying antibiotic resistance genes, as well as mobile genetic elements were investigated in manured soil in several previous studies [17, 19, 32, 60]. Manure amendment was reported to enhance transfer of antibiotic resistance genes in soil [8, 24, 34]. From these studies, it became clear that the fate and effects of bacteria and veterinary medicines entering with manure into the soil needs to be better understood. Mobile genetic elements carrying antibiotic resistance genes can be easily transferred to bacteria associated with humans through ground water or food [12, 21, 46, 55, 64]. Consequently, it cannot be excluded that this will affect human and animal health.

Acknowledgments

This study was funded by the DFG research group FOR566 "Veterinary Medicines in Soils". We are grateful to H. Döhler for organizing the collection of manure samples. We acknowledge the help by E. Krögerrecklenfort and N. Mrotzek. Furthermore, we thank Ms I.-M. Jungkurth for critically reading this manuscript. C.T.T. Binh from Vietnam was supported by the DAAD.

References

[1] Agersø, Y. & Sandvang, D. (2005). Class 1 integrons and tetracycline resistance genes in *Alcaligenes*, *Arthrobacter*, and *Pseudomonas* spp. isolated from pigsties and manured soil. *Appl. Environ. Microbiol.*, *71*, 7941-7947.

[2] Alexander, C. J., Citron, D. M., Brazier, J. S. & Goldstein, E. J. C. (1995). Identification and antimicrobial resistance pattern of clinical isolates of *Clostridium clostridioforme*, *Clostridium innocuum,* and *Clostridium ranosum* compared with those of clinical isolates of *Clostridium perfringens*. *J. Clin. Microbiol.*, *33*, 3209-3215.

[3] Antunes, P., Machado, J. & Peixe, L. (2007). Dissemination of *sul3*-containing elements linked to class 1 integrons with an unusual 3' conserved sequence region among *Salmonella* isolates. *Antimicrob. Agents Chemother.*, *51*, 1545-1548.

[4] Antunes, P., Machado, J., Sousa, J. C. & Peixe, L. (2005). Dissemination of sulfonamide resistance genes (*sul1*, *sul2*, and *sul3*) in Portuguese *Salmonella enterica* strains and relation with integrons. *Antimicrob. Agents Chemother.*, *49*, 836-839.

[5] Baloda, S. B., Christensen, L. & Trajcevska, S. (2001). Persistence of a *Salmonella enterica* serovar Typhimurium DT12 clone in a piggery and

in agricultural soil amended with *Salmonella*-contaminated slurry. *Appl. Environ. Microbiol.*, *67*, 2859-2862.

[6] Bendle, J. S., James, P. A., Bennett, P. M., Avison, M. B., MacGowan, A. P. & Al-Shafi, K. M. (2004). Resistance determinants in strains of *Clostridium difficile* from two geographically distinct populations. *Int. J. Antimicrob. Agents*, *24*, 619-621.

[7] Bibbal, D., Dupouy, V., Ferré, J. P., Toutain, P. L., Fayet, O., Prère, M. F. & Bousquet-Mélou, A. (2007). Impact of three ampicillin dosage regimens on selection of ampicillin resistance in *Enterobacteriaceae* and excretion of bla_{TEM} genes in swine feces. *Appl. Environ. Microbiol.*, *73*, 4785-4790.

[8] Binh, C. T., Heuer, H., Gomes, N. C. M., Kotzerke, A., Fulle, M., Wilke, B. M., Schloter, M. & Smalla, K. (2007). Short-term effects of amoxicillin on bacterial communities in manured soil. *FEMS Microbiol. Ecol.*, *62*, 290-302.

[9] Binh, C. T., Heuer, H., Kaupenjohann, M. & Smalla, K. (2008). Piggery manure used for soil fertilization is a reservoir for transferable antibiotic resistance plasmids. *FEMS Microbiol. Ecol.*, *66*, 25-37.

[10] Binh, C. T., Heuer, H., Kaupenjohann, M. & Smalla, K. (2009). Diverse *aadA* gene cassettes on class 1 integrons introduced into soil via spread manure. *Res. Microbiol.*, *160*, 427-433.

[11] Bush, K. (1997) In: D. J. Chadwick, J. Goode, (Eds.), Antibiotic resistance: origins, evolution, selection and spread, John Wiley & Sons, Chichester, 152-163.

[12] Chee-Sanford, J. C., Aminov, R. I., Krapac, I. J., Garrigues-Jeanjean, N. & Mackie, R. I. (2001). Occurrence and diversity of tetracycline resistance genes in lagoons and groundwater underlying two swine production facilities. *Appl. Environ. Microbiol.*, *67*, 1494-1502.

[13] Chen, J., Yu, Z., Michel, F. C., Wittum, T. & Morrison, M. (2007). Development and application of real-time PCR assays for quantification of *erm* genes conferring resistance to macrolides-lincosamides-streptogramin B in livestock manure and manure management systems. *Appl. Environ. Microbiol.*, *73*, 4407-4416.

[14] Colom, K., Pérez, J., Alonso, R., Fernández-Aranguiz, A., Lariño, E. & Cisterna, R. (2003). Simple and reliable multiplex PCR assay for detection of bla_{TEM}, bla_{SHV}, $bla_{\text{OXA-1}}$ genes in *Enterobacteriaceae*. *FEMS Microbiol. Lett.*, *223*, 147-151.

[15] Cotta, M., Whitehead, T. R. & Zeltwanger, R. L. (2003). Isolation, characterization and comparison of bacteria from swine faeces and

manure storage pits. *Environ. Microbiol.*, *5*, 737-745.

[16] David, V., Bozdogan, B., Mainardi, J. L., Legrand, R., Gutmann, L. & Leclerq, R. (2004). Mechanism of intrinsic resistance to vancomycin in *Clostridium innocuum* NCIB 10674. *J. Bacteriol.*, *186*, 3415-3422.

[17] Drønen, A. K., Torsvik, V. & Top, E. M. (1999). Comparison of the plasmid types obtained by two distantly related recipients in biparental exogenous plasmid isolations from soil. *FEMS Microbiol. Lett.*, *176*, 105-110.

[18] Duriez, P. & Topp, E. (2007). Temporal dynamics and impact of manure storage on antibiotic resistance patterns and population structure of *Escherichia coli* isolates from a commercial swine farm. *Appl. Environ. Microbiol.*, *73*, 5486-5493.

[19] Elsas, J. D., Gardener, B. B. M., Wolters, A. C. & Smit, E. (1998). Isolation, characterization, and transfer of cryptic gene-mobilizing plasmids in the wheat rhizosphere. *Appl. Environ. Microbiol.*, *64*, 880-889.

[20] Enne, V. I., King, A., Livermore, D. M. & Hall, L. M. C. (2002). Sulfonamide resistance in *Haemophilus influenzae* mediated by acquisition of *sul2* or a short insertion in chromosomal *folP*. *Antimicrob. Agents Chemother.*, *46*, 1934-1939.

[21] Espinosa-Urgel, M. (2004). Plant-associated *Pseudomonas* populations: molecular biology, DNA dynamics, and gene transfer. *Plasmid*, *52*, 139-150.

[22] Fremaux, B., Pringent-Combaret, C., Delignette-Muller, M. L., Mallen, B., Dothal, M., Gleizal, A. & Vernozy-Rozand, C. (2008). Persistence of Shiga toxin-producing *Escherichia coli* O26 in various manure-amended soil types. *J. Appl. Microbiol.*, *104*, 296-304.

[23] Gomes, N. C. M., Borges, L. R., Paranhos, R., Pinto, F. N., Mendonça-Hagler, L. C. S. & Smalla, K. (2008). Exploring the diversity of bacterial communities in sediments of urban mangrove forests. *FEMS Microbiol. Ecol.*, *66*, 96-109.

[24] Götz, A. & Smalla, K. (1997). Manure enhances plasmid mobilization and survival of *Pseudomonas putida* introduced into field soil. *Appl. Environ. Microbiol.*, *63*, 1980-1986.

[25] Guerra, B., Junker, E. & Helmuth, R. (2004). Incidence of the recently described sulfonamide resistance gene *sul3* among German *Salmonella enterica* strains isolated from livestock and food. *Antimicrob. Agents Chemother.*, *48*, 2712-2715.

[26] Guerra, B., Junker, E., Schroeter, A., Malorny, B., Lehmann, S. &

Helmuth, R. (2003). Phenotypic and genotypic characterization of antimicrobial resistance in German *Escherichia coli* isolates from cattle, swine and poultry. *J. Antimicrob. Chemother.*, *52*, 489-492.

[27] Halling-Sørensen, B., Nors, N. S., Lanzky, P. F., Ingerslev, F., Lutzhøft, H. C. H. & Jørgensen, S. E. (1998). Occurrence, fate and effects of pharmaceutical substances in the environment - a review. *Chemosphere*, *36*, 357-393.

[28] Henriques, I. S., Fonseca, F., Alves, A., Saavedra, M. J. & Correia, A. (2006). Analysing diversity among β-lactamase encoding genes in aquatic environments. *FEMS Microbiol. Ecol.*, *56*, 418-429.

[29] Henriques, I. S., Fonseca, F., Alves, A., Saavedra, M. J. & Correia, A. (2006). Occurrence and diversity of integrons and β-lactamase genes among ampicillin-resistant isolates from estuarine waters. *Res. Microbiol.*, *157*, 938-947.

[30] Heuer, H., Focks, A., Lamshöft, M., Smalla, K., Matthies, M. & Spiteller, M. (2008). Fate of sulfadiazine administered to pigs and its quantitative effect on the dynamics of bacterial resistance genes in manure and manured soil. *Soil Biol. Biochem.*, *40*, 1892-1900.

[31] Heuer, H., Kopmann, C., Binh, C.T., Top, E. M. & Smalla, K. (2009). Spreading antibiotic resistance through spread manure: characteristics of a novel plasmid type with low %G+C content. *Environ. Microbiol.*, *11*, 937-949.

[32] Heuer, H., Krögerrecklenfort, E., Wellington, E. M. H., Egan, S., van Elsas, J. D., van Overbeek, L., Collard, J. M., Guillaume, G., Karagouni, A. D., Nikolakopoulou, T. L. & Smalla, K. (2002). Gentamicin resistance genes in environmental bacteria: prevalence and transfer. *FEMS Microbiol. Ecol.*, *42*, 289-302.

[33] Heuer, H., Krsek, M., Baker, P., Smalla, K. & Wellington, E. M. H. (1997). Analysis of actinomycete communities by specific amplification of genes encoding 16S rRNA and gel-electrophoretic separation in denaturing gradients. *Appl. Environ. Microbiol.*, *63*, 3233-3241.

[34] Heuer, H. & Smalla, K. (2007). Manure and sulfadiazine synergistically increased bacterial antibiotic resistance in soil over at least two months. *Environ. Microbiol.*, *9*, 657-666.

[35] Heuer, H., Wieland, G., Schönfeld, J., Schönwälder, A., Gomes, N. C. M. & Smalla, K. (2001). Environmental Molecular Microbiology: Protocols and Applications, Horizon Scientific Press, Wymondham, UK, 178-190

[36] Jindal, A., Korcherginskaya, S., Mehboob, A., Robert, M., Mackie, R. I.,

Raskin, L. & Zilles, J. L. (2006). Antimicrobial use and resistance in swine waste treatment systems. *Appl. Environ. Microbiol.*, *72*, 7813-7820.

[37] Kather, E. J., Marks, S. L. & Foley, J. E. (2006). Determination of the prevalence of antimicrobial resistance genes in canine *Clostridium perfringens* isolates. *Vet. Microbiol.*, *113*, 97-101.

[38] Kotzerke, A., Sharma, S., Schauss, K., Heuer, H., Thiele-Bruhn, S., Smalla, K., Wilke, B. M. & Schloter, M. (2008). Alterations in soil microbial activity and N-transformation processes due to sulfadiazine loads in pig-manure. *Environmental Pollution*, *153*, 315-322.

[39] Kümmerer, K. (2003). Significance of antibiotics in the environment. *J. Antimicrob. Chemother.*, *52*, 5-7.

[40] Langford, F. M., Weary, D. M. & Fisher, L. (2003). Antibiotic resistance in gut bacteria from dairy calves: a dose response to the level of antibiotics fed in milk. *J. Dairy Sci.*, *86*, 3963-3966.

[41] Livermore, D. M. (1995). β-lactamases in laboratory and clinical resistance. *Clin. Microbiol. Rev.*, *8*, 557-584.

[42] Ludwig, W., Strunk, O. & Westram, R. (2004). ARB: a software environment for sequence data. *Nucleic Acids Res.*, *32*, 1363-1371.

[43] Mathew, A. G., Cissel, R. & Liamthong, S. (2007). Antibiotic resistance in bacteria associated with food animals: a United States perspective of livestock production. *Foodborne Pathog. Dis.*, *4*, 115-133.

[44] Mathew, A. G., Upchurch, W. G. & Chattin, S. E. (1998). Incidence of antibiotic resistance in fecal *Escherichia coli* isolated from commercial swine farms. *J. Anim. Sci.*, *76*, 429-434.

[45] Melville, C. M., Scott, K. P., Mercer, D. K. & Flint, H. J. (2001). Novel tetracycline resistance gene, *tet*(32), in the *Clostridium*-related human colonic anaerobe K10 and its transmission *in vitro* to the rumen anaerobe *Butyrivibrio fibrisolvens*. *Antimicrob. Agents Chemother.*, *45*, 3246-3249.

[46] Mølbak, L., Licht, T. R., Kvist, T., Kroer, N. & Andersen, S. R. (2003). Plasmid transfer from *Pseudomonas putida* to the indigenous bacteria on alfalfa sprouts: characterization, direct quantification, and *in situ* location of transconjugant cells. *Appl. Environ. Microbiol.*, *69*, 5536-5542.

[47] Nagai, K., Davies, T. A., Jacobs, M. R. & Appelbaum, P. C. (2002). Effects of amino acid alterations in penicillin-binding proteins (PBPs) 1a, 2b, and 2x on PBP affinities of penicillin, ampicillin, amoxicillin, cefditoren, cefuroxime, cefprozil, and cefaclor in 18 clinical isolates of

penicillin-susceptible, -intermediate, and -resistant pneumococci. *Antimicrob. Agents Chemother.*, *46*, 1273-1280.

[48] Neikirk, R. L. & Krieg, N. R. (1964). Sulfonamide resistance of *Clostridium sporogenes*. *J. Bacteriol.*, *87*, 1526-1527.

[49] Overbeek, L. S., Wellington, E. M. H., Egan, S., Smalla, K., Heuer, H., Collard, J. M., Guillaume, G., Karagouni, A. D., Nikolakopoulou, T. L. & Elsas, J. D. (2002). Prevalence of streptomycin-resistance genes in bacterial populations in European habitats. *FEMS Microbiol. Ecol.*, *42*, 277-288.

[50] Pei, R., Kim, S. C., Carlson, K. H. & Pruden, A. (2006). Effect of river landscape on the sediment concentrations of antibiotics and corresponding antibiotic resistance genes (ARG). *Water Res.*, *40*, 2427-2435.

[51] Perreten, V. & Boerlin, P. (2003). A new sulfonamide resistance gene (*sul3*) in *Escherichia coli* is widespread in the pig population of Switzerland. *Antimicrob. Agents Chemother.*, *47*, 1169-1172.

[52] Peu, P., Brugère, H., Pourcher, A. M., Kérourédan, M., Godon, J. J., Delgenès, J. P. & Dabert, P. (2006). Dynamics of a pig slurry microbial community during anaerobic storage and management. *Appl. Environ. Microbiol.*, *72*, 3578-3585.

[53] Rådström, P. & Swedberg, G. (1988). RSF1010 and a conjugative plasmid contain *sul2*, one of the two known genes for plasmid-borne sulfonamide resistance dihydropteroate synthase. *Antimicrob. Agents Chemother.*, *32*, 1684-1692.

[54] Rådström, P., Swedberg, G. & Sköld, O. (1991). Genetic analyses of sulfonamide resistance and its dissemination in Gram-negative bacteria illustrate new aspects of R plasmid evolution. *Antimicrob. Agents Chemother.*, *35*, 1840-1848.

[55] Salyers, A. A. (2002). An overview of the genetic basis of antibiotic resistance in bacteria and its implications for agriculture. *Animal Biotechnol.*, *13*, 1-5.

[56] Schweizer, G., Ehrensperger, F., Torgerson, P. R. & Braun, U. (2006). Clinical findings and treatment of 94 cattle presumptively diagnosed with listeriosis. *Vet. Rec.*, *158*, 588-592.

[57] Sengeløv, G., Agersø, Y., Halling-Sørensen, B., Baloda, S. B., Andersen, J. S. & Jensen, L. B. (2003). Bacterial antibiotic resistance levels in Danish farmland as a result of treatment with pig manure slurry. *Environ. Int.*, *28*, 587-595.

[58] Sköld, O. (2001). Resistance to trimethoprim and sulfonamides. *Vet.*

Res., *32*, 261-273.

[59] Smalla, K., Heuer, H., Götz, A., Niemeyer, D., Krögerrecklenfort, E. & Tietze, E. (2000). Exogenous isolation of antibiotic resistance plasmids from piggery manure slurries reveals a high prevalence and diversity of IncQ-like plasmids. *Appl. Environ. Microbiol.*, *66*, 4854-4862.

[60] Smalla, K. & Sobecky, P. (2002). The prevalence and diversity of mobile genetic elements in bacterial communities of different environmental habitats: insights gained from different methodological approaches. *FEMS Microbiol. Ecol.*, *42*, 165-175.

[61] Spigaglia, P., Barbanti, F. & Mastrantonio, P. (2006). New variants of the *tet*(M) gene in *Clostridium difficile* clinical isolates harbouring Tn*916*-like elements. *J. Antimicrob. Chemother.*, *57*, 1205-1209.

[62] Sundström, L., Rådström, P., Swedberg, G. & Sköld, O. (1988). Site-specific recombination promotes linkage between trimethoprim- and sulfonamide-resistance genes: sequence characterization of *dfr5* and *sul1* and a recombination active locus of Tn*21*. *Mol. Gen. Genet.*, *213*, 191-201.

[63] Takai, K. & Horikoshi, K. (2000). Rapid detection and quantification of members of the archaeal community by quantitative PCR using fluorogenic probes. *Appl. Environ. Microbiol.*, *66*, 5066-5072.

[64] Teuber, M. (2001). Veterinary use and antibiotic resistance. *Curr. Opin. Microbiol.*, *4*, 493-499.

[65] Whitehead, T. R. & Cotta, M. A. (2001). Characterization and comparison of microbial populations in swine faeces and manure storage pits by 16S rDNA gene sequence analyses. *Anaerobe*, *7*, 181-187.

[66] Witte, W. (2000). Selective pressure by antibiotic use in livestock. *Int. J. Antimicrob. Agents*, *16*, S19-S24.

[67] Yu, Z., Michel, F. C., Hansen, G., Wittum, T. & Morrison, M. (2005). Development and application of real-time PCR assays for quantification of genes encoding tetracycline resistance. *Appl. Environ. Microbiol.*, *71*, 6926-6933.

In: Manure: Management, Uses and… ISBN: 978-1-61668-424-2
Editors: C. S. Dellaguardia, pp. 171-184

Chapter 6

BIOREMEDIATIONAL TECHNIQUE OF ADDING NITRITE-OXIDIZING BACTERIA FOR REDUCING N_2O EMISSION DURING SWINE MANURE COMPOSTING

Yasuyuki Fukumoto*

Pollution Control Research Team, National Institute of Livestock & Grassland Science, 2 Ikenodai, Tsukuba, Ibaraki 305-0901, Japan

ABSTRACT

Livestock waste treatment holds a large nitrous oxide (N_2O) potential with global emission. Composting is a principal treatment method of solid fraction of swine manure in Japan. However, swine manure composting entails substantial emissions of nitrogenous compounds including N_2O. During swine manure composting, N_2O is mainly emitted after thermophilic phase because the nitrification mediated by mesophilic microorganisms (ammonia-oxidizing bacteria [AOB] and nitrite-oxidizing bacteria [NOB]) is a necessary prerequisite for N_2O generation from fresh manure which contains little nitrate (NO_3^-) nitrogen. However, the growth of indigenous NOB after thermophilic phase tends to be late than that of AOB, which could be a cause of toxic nitrite (NO_2^-) accumulation during maturation phase of composting. It is known that

* Corresponding author: E-mail: yasuyuki@affec.go.jp.

there is a strong correlation between N_2O emission and NO_2^- accumulation in the composting process; therefore, the NO_2^- accumulation after thermophilic phase induces substantial N_2O emission. For the purpose of shortening the time lag of proliferation between AOB and NOB, a bioremediational technique of adding mature compost, including NOB, at the start of AOB proliferation was examined in the experiments of swine manure composting. The adding mature compost after thermophilic phase made it possible to establish NOB guild more quickly in the composting material, which could prevent NO_2^- accumulation by prompt oxidization from NO_2^- to NO_3^-. As the pattern of N_2O emission agreed well with that of NO_2^- concentration, N_2O emission ceased rapidly when NOB was added. Summarizing data of laboratory-scale experiments, the average reducing rate of N_2O emission by the mature compost (NOB source) addition from the control was calculated to be 60%, and therefore, this technique appeared to allow a quantitative reduction in N_2O emission in the swine manure composing. From the result of nitrogen mass balance in the composting experiment, it was found that the NO_2^- accumulation induced not only N_2O, but also other nitrogenous emissions during compost maturation; therefore, a possibility that if NO_2^- accumulation has been prolonged the nitrogen loss induced by NO_2^- accumulation could become a principal cause of the total nitrogen losses was suggested. To reduce such a serious environmental risk as global warming and improve the efficiency of nitrogen from animal compost, the bioremediational technique for complete nitrification promotion is considered to be effective for mitigation of N_2O emission from swine manure composting.

INTRODUCTION

The amount of animal wastes excreted per year in Japan was 8,747 million ton in 2008 (Ministry of Agriculture, Forestry and Fisheries, Japan, 2008), and it is estimated that approximately 0.7 million ton of nitrogen is contained in those excretion, which is almost equal to the total amount of nitrogen annually consumed as chemical fertilizer in Japan [9]. Therefore, it is important to improve the use efficiency of nitrogen contained in animal wastes for sustainable agricultural systems.

In many swine husbandry in Japan, it is indispensable to treat swine excretions because the most of swine farmers do not have large farmland capable to apply huge manure. Usually, the feces, urine and washing water from piggeries are separated mechanically into solid and liquid fractions. The liquid fraction is discharged into public waters after the purification, and the solid fraction is composted and utilized for farmland fertilizer.

Composting is a general treatment method of animal wastes in Japan. In the composting process, large parts of easy-digestible organic matter, which is a cause of unsanitary properties of animal wastes, are decomposed aerobically by the activity of different microorganisms. After the composting, the odoriferous and unsanitary animal wastes are able to convert into odorless, sanitary and marketable organic fertilizer. However, it is well known that composting entails substantial emissions of harmful gaseous components [15, 17].

Because of high nitrogen content in animal wastes, large amounts of ammonium (NH_4^+) are produced in the early degradation stage of composting, and an alkaline pH condition during thermophilic phase induces substantial losses of gaseous ammonia (NH_3). NH_3 emission is causes of not only several environmental hazards such as odor problem and acid rain [18], but also diminishing the value of final product as a fertilizer [2, 14]. The production of NH_4^+ and NH_3 during composting would stimulate nitrification, a two-step microbial process consisting of the oxidation of NH_3 to nitrite (NO_2^-) and its subsequent oxidation to nitrate (NO_3^-). The kinds of gaseous nitrogenous compound change suddenly before and after nitrification start during composting. After nitrification start in the swine manure composting, the substantial NH_3 emission ceases quickly; however, instead of that, a significant nitrous oxide (N_2O) emission is generally begun [6].

N_2O is an important atmospheric component, behaves as a strong greenhouse gas with 298 times more global warming potential than CO_2 over a 100 years time horizon [12]. In addition, N_2O reacts with oxygen to form nitric oxide in the stratosphere, which plays a role in the deterioration of the ozone layer [3]. In Japan, the amount of N_2O emission from livestock wastes treatments is estimated at 4.9 million ton (CO_2 equivalent) in 2007 (Greenhouse Gas Inventory Office of Japan), which holds 20% of the total nation's N_2O emissions. In the world, the livestock activities contribute two-third of anthropogenic N_2O emissions, most of which would be due to the treatment and utilization of livestock manure [4]. Moreover, because of the fast growth of global livestock industry, it is expected that the contribution to the global N_2O emissions will substantially increase over the coming decade. These backgrounds indicate the importance of N_2O mitigation from livestock industry. However, especially in the treatment of animal wastes, there are few effective and feasible countermeasures at present. The subject of this chapter is to show one possibility of N_2O mitigation which has been found in swine manure composting.

NITRIFICATION AND N_2O EMISSION DURING SWINE MANURE COMPOSTING

Nitrification is a necessary prerequisite for N_2O generation from livestock manure because little nitrites and nitrates are contained in the fresh excretions. Generally, during swine manure composting, obvious nitrification activity begins after thermophilic phase in which temperature of composting material rises up to over 60 °C. This is because the chemoautotrophic nitrifying bacteria, which play the central role in nitrification in composting, prefer a mesophilic temperature range of 20 to 30 °C [16]. Moreover, high free ammonia concentration, low oxygen level, and competition with heterotrophic microorganisms during thermophilic phase may become negative factors for the growth of nitrifying bacteria [1].

Chemolithotrophic nitrification is performed by two groups of bacteria. One is ammonia-oxidizing bacteria (AOB), represented by the genera *Nitrosomonas* and *Nitrosospira*, oxidizing NH_3 to NO_2^-, and the another is nitrite-oxidizing bacteria, represented by the genera *Nitrobacter* and *Nitrospira*, oxidizing NO_2^- to NO_3^-. For the complete nitrification, the establishment of well balanced community of the both AOB and NOB guilds is necessary. During swine manure composting, however, it seems difficult to establish such a balanced nitrifying community sooner by indigenous AOB and NOB.

When the condition of composting material has been changed to allow nitrifying bacteria's activities after thermophilic phase, AOB tends to proliferate solely, and they begin to abruptly oxidize NH_3. However, the growth of indigenous NOB tends to be delayed greatly compared with that of AOB, which causes NO_2^- accumulation for a while after thermophilic phase of swine manure composting. It has long been understood that there is a strong correlation between N_2O emission and NO_2^- accumulation in the composting process [11]; therefore, the amount of N_2O emission would increase with prolonging duration of NO_2^- accumulation. That is, it appears that to prevent NO_2^- accumulation becomes a countermeasure for reducing N_2O emission during swine manure composting.

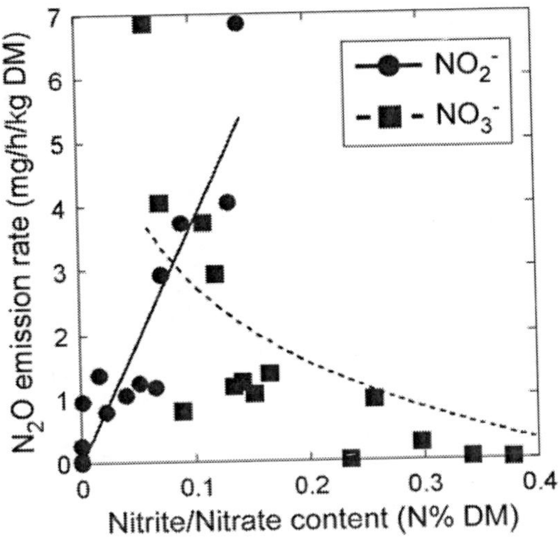

Figure 1. Correlation between N_2O emission rate and nitrite/nitrate content in swine manure

There are two ways of thinking to prevent NO_2^- accumulation during composting process. One idea is to use reagent, nitrification inhibitors. Because nitrification inhibitors prevent AOB from oxidizing NH_3, it seems to be possible to prevent NO_2^- accumulation and reduce N_2O emission as a result. However, the utilization of nitrification inhibitors during composting holds some problems in practice. The biggest problem is that a new cost about the reagents occurs in the animal wastes treatment, which will suppress the management of livestock farmers. Moreover, there is a possibility that the environmental problems with NH_3 emission become intensified because of the accumulation of NH_4^+. Therefore, the utilization of nitrification inhibitors for reducing N_2O emission seems infeasible in the actual treatments.

While NO_2^- concentration is direct proportional to N_2O emission rate, it was found that there is a negative correlation between N_2O and NO_3^- during swine manure composting (Figure 1). This indicated the possibility of reducing N_2O emission by prompt oxidization from NO_2^- to NO_3^-. As shown above, one of the major causes of NO_2^- accumulation in the swine manure composting is the delayed growth of indigenous NOB compared with that of AOB. Therefore, a bioremediational technique of adding NOB at the start of nitrification seemed to become an effective countermeasure for reducing N_2O emission from swine manure composting.

BIOREMEDIATIONAL TECHNIQUE FOR REDUCING N_2O EMISSION

To quantify the effect of NOB addition on N_2O emission, swine manure composting experiments were conducted using the laboratory-scale composting apparatuses [8]. As the source of NOB, mature swine compost was used in a series of composting experiments. The reason of having selected mature swine compost was that farmers can get it easily without any cost. Moreover, it was also expected that the genera composition of nitrifying bacteria built in mature swine compost would be fit for nitrification progress in the composting of same kind of animal feces. It is difficult to define the "mature" swine compost generally. However, in this case, it was defined from a viewpoint of nitrogen turnover, i.e., whether the complete nitrification is achieved. When the complete nitrification was achieved, NOB has been contained in mature swine compost at approximately 10^5-10^6 cells per gram of compost in a series of composting experiments. During the composting treatment, mature swine compost was added after thermophilic phase to avoid decreasing the number of NOB added at high temperature.

By the addition of mature swine compost, NOB guild has been immediately built in the composting material, which could contribute to complete nitrification promotion. On the other hand, in our past experiments, a significant time lag of proliferation between AOB and NOB was almost confirmed in the experimental run without NOB addition (control). From those results, it has become obvious that the unbalanced composition of nitrifying communities can be modified to normal state by the artificial addition of NOB during swine manure composting (Figure 2).

The transition of inorganic nitrogen compounds has been reflected by the growth style of nitrifying bacteria. After thermophilic phase of swine manure composting, the prolonged accumulations of NO_2^- occurred in the control, because of the delayed growth of the indigenous NOB. On the other hand, the addition of mature swine compost, containing NOB, could contribute to prompt oxidization from NO_2^- to NO_3^-, and prevention of prolonged NO_2^- accumulation. Moreover, the amount of NO_3^- preserved in final product was obviously increased by the addition of NOB (Figure 3).

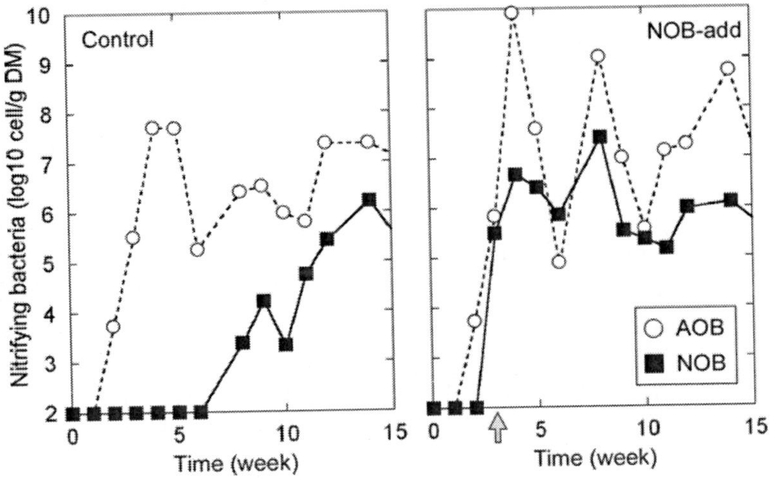

Figure 2. Temporal changes in the number of nitrifying bacteria during swine manure composting. Arrow indicates the addition of mature swine compost. AOB, ammonia-oxidizing bacteria; NOB, nitrite-oxidizing bacteria. Detection limit is less than 10^2 cell/g material

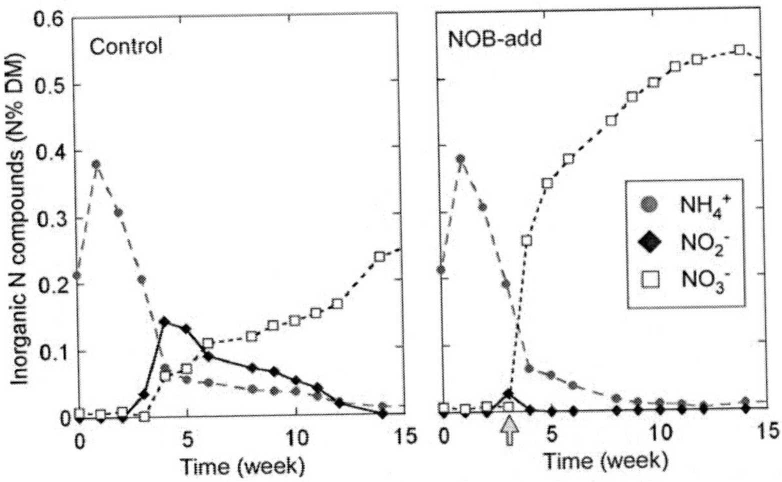

Figure 3. Temporal changes in the concentrations of inorganic nitrogen compounds during swine manure composting. Arrow indicates the addition of mature swine compost

The patterns of N_2O emission have reflected the trend of NO_2^- accumulation. During thermophilic phase, N_2O emission shows a low level change because of the low nitrification activity. When the fermentation stage changes from thermophilic to mesophilic, N_2O emission forms a sharp peak with the abrupt oxidization of NH_3 by proliferated AOB. After the peak formation, N_2O emission began to decline; however, in the control, it was kept at higher concentration than the atmospheric level during the almost same duration of NO_2^- accumulation. On the other hand, N_2O emission ceased rapidly after the peak formation in the treatments with NOB addition, which would be due to the short duration of NO_2^- accumulation. Because of shortening the significant emission period, the amount of total N_2O emission during composting was also decreased compared with the control. Through a series of our experiments in past, the average reducing rate of N_2O emission was calculated to be 61.3% (Figure 4). From these results, it is thought that this technique allows a quantitative reduction in N_2O emission during swine manure composting.

EFFECT OF NITRITE ACCUMULATION ON TOTAL NITROGEN LOSS

The amount of N_2O emission during swine manure composting can be reduced by NOB addition, and therefore, nitrogen content in the final product is increased, which may contribute to raise the value as a fertilizer. Actually, the amount of NO_3^--N has been increased in the treatment of NOB addition. However, it is generally thought that the major part of nitrogen losses in composting is due to the NH_3 emission during thermophilic phase, which suggests that the effect of N_2O reduction on nitrogen preservation is not so great. Thereupon, the experiment has been conducted to investigate the effect of NOB addition on nitrogen preservation during swine manure composting [5]. In that experiment, the emission rate of N_2O was decreased by NOB addition from 9.27% to 2.97% of the initial total nitrogen ($TN_{initial}$), while there was little difference in the NH_3 emission rate (10% of $TN_{initial}$). Therefore, the nitrogen loss has been mitigated by 33% as far as NH_3 and N_2O emissions are concerned. However, the difference in the nitrogen loss was markedly magnified as a result of nitrogen balance study, which were 27.8% (control) and 13.3% (NOB addition) of $TN_{initial}$, respectively, mitigating nitrogen losses by 52% as the whole (Figure 5). This result showed that the NO_2^- accumulation would induce not only N_2O, but also other nitrogenous emissions during the composting process. Therefore, the treatment of NOB addition may have a greater effect on nitrogen preservation than expected.

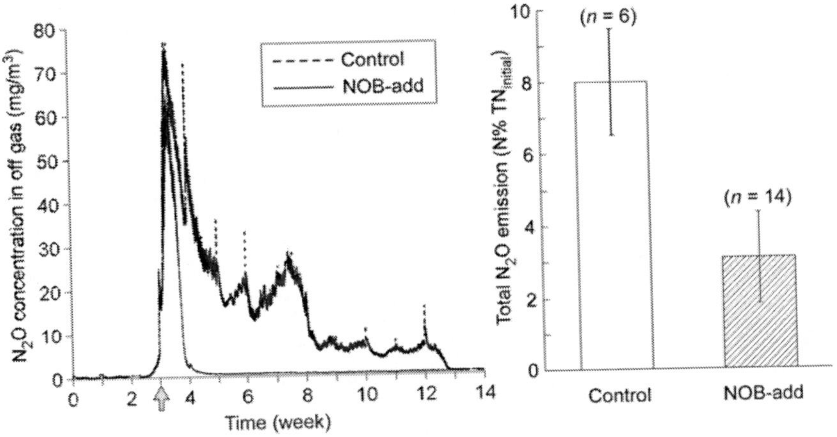

Figure 4. Effect of nitrite-oxidizing bacteria addition on the emission patterns (left) and total emissions (right) of N_2O during swine manure composting. Arrow indicates the addition of mature swine compost, and error bars indicate standard deviation.

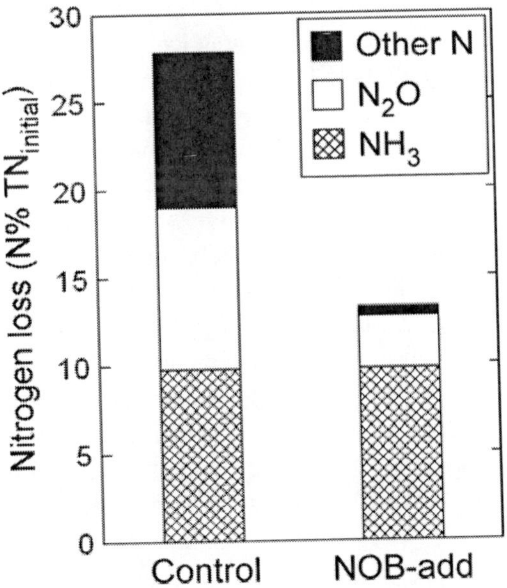

Figure 5. Details of nitrogen losses during swine manure composting.

From the above-mentioned result, a new question surfaced. That is, what kinds of components have consisted in the unknown parts of nitrogen losses? To reveal the components, nitric oxide (NO) was measured in the later composting experiment because some studies showed that the NO can be produced under aerobic conditions in the presence of NO_2^- [19, 20]. As the result, it was confirmed that NO is one of the components consisting in that unknown parts (data not shown). Kester *et al.* [13] has already reported a similar result that the NO_2^- accumulation enhanced both N_2O and NO emissions in the continuous culture of nitrifiers and denitrifiers with aeration. NO is a precursor of tropospheric ozone, which is a greenhouse gas by photochemical reaction. Moreover, NO_2 is derived from NO and is a precursor of nitric acid, which is a major component of acid rain. Further study will be necessary to evaluate the NO emission from a quantitative viewpoint.

Issues for Practical Use

The bioremediational technique shown in present study is thought to be effective to reduce N_2O and other nitrogenous emissions induced by NO_2^- accumulation during swine manure composting (Figure 6). However, before this technique is established to allow practical use, some problems must be overcome. The first problem is how to decide the timing of NOB addition during composting process. In a series of composting experiments using laboratory-scale apparatus, the temperature of compost material was one of the indicators to decide the timing of NOB addition. However, in actual treatment, the temperature profile inside a compost pile is not always uniform [7, 10]. For example, the temperature of surface position declines earlier than that inside the pile, which would make partial differences in the growth rate of nitrifiers inside the compost pile. Therefore, there is a possibility to miss the opportunity of NOB addition when the temperature is used as an only index. And moreover, another problem is that these targeted gases do not have pungency such as NH_3, which makes it difficult to confirm the effect of treatment. In the forthcoming study, the composting experiment close to actual treatment should be conducted to solve those problems and extract other new problems.

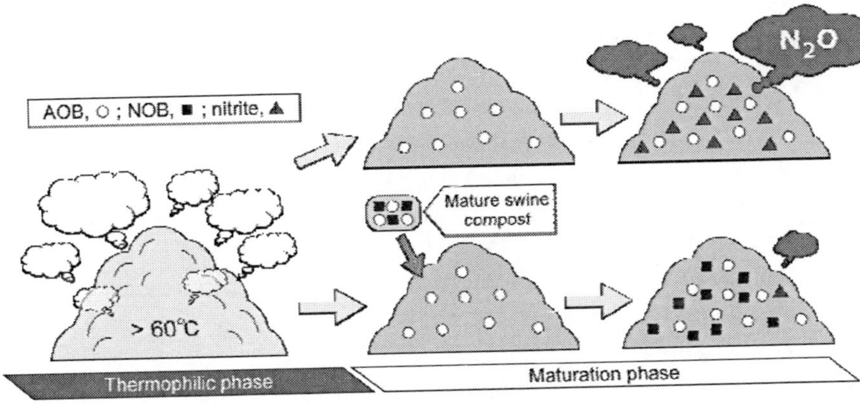

Figure 6. Schematic of the bioremediational technique for reducing N_2O emission during swine manure composting. AOB, ammonia-oxidizing bacteria; NOB, nitrite-oxidizing bacteria

CONCLUSION

During swine manure composting, NO_2^- accumulation caused by the delayed growth of indigenous NOB induces nitrogenous emissions such as N_2O enhancing global environmental risk. A bioremediational technique of adding NOB (contained in mature swine compost) prevented prolonged NO_2^- accumulation by prompt oxidization from NO_2^- to NO_3^-, and as a result, the amount of N_2O emission could be reduced by 60% as the average in a series of laboratory-scale experiments. From the result of nitrogen balance study, it was found that the NO_2^- accumulation would induce not only N_2O, but also other nitrogenous emissions in which NO is contained as a component, which suggests the production possibility of compost with higher nitrogen contents by prevention of NO_2^- accumulation. To reduce atmospheric pollutions and raise the value of compost as a fertilizer, the technique shown in this chapter is thought to become one of the effective countermeasures in the composting of swine manure.

Acknowledgements

This work was supported in part by a Grant-in-Aid for Scientific Research (20780235) from the Ministry of Education, Culture, Sports, Science and Technology of Japan, and by a Rural Biomass Research Project, BM-D3120 from the Ministry of Agriculture, Forestry and Fisheries of Japan. The author wishes to thank Dr. K. Haga, Dr. K. Suzuki, Dr. T. Osada, Mr. K. Kuroda, Dr. D. Hanajima, Ms. T. Yasuda, and Prof. K. Inubushi for their scientific advice and valuable help for composting experiments.

Parts of this study have been published in Ref. 5 & 8.

Abbreviations

AOB	ammonia-oxidizing bacteria
NOB	nitrite-oxidizing bacteria
TN	total nitrogen
DM	dry matter

References

[68] Anthonisen, A. C., Loehr, R. C., Prakasam, T. B. S. & Srinath, E. G. (1976). Inhibition of nitrification by ammonia and nitrous-acid. *J. Water Pollut. Control Fed.*, *48*, 835-852.

[69] Beck-Friis, B., Smårs, S., Jönsson, H. & Kirchmann, H. (2001). Gaseous emissions of carbon dioxide, ammonia and nitrous oxide from organic household waste in a compost reactor under different temperature regimes. *J. Agric. Eng. Res.*, *78*, 423-430.

[70] Crutzen, P. J. (1981). Atmospheric chemical processes of the oxides of nitrogen, including nitrous oxide. In: *Denitrification, Nitrification and Atmospheric Nitrous Oxide*, C. C. Delwiche, (Ed.), John Wiley and Sons, New York, 17-44.

[71] FAO, (2006). Livestock's role in climate change and air pollution. In: *Livestock's long shadow*, H., Steinfeld, P., Gerber, T., Wassenaar, V., Castel, M. Rosales, & C. de Haan (Eds.), *LEAD Virtual Research and Development Centre*, Rome, Italy, 102-123.

[72] Fukumoto, Y. & Inubushi, K. (2009). Effect of nitrite accumulation on

nitrous oxide emission and total nitrogen loss during swine manure composting. *Soil Sci. Plant Nutr.*, *55*, 428-434.

[73] Fukumoto, Y., Osada, T., Hanajima, D. & Haga, K. (2003). Patterns and quantities of NH_3, N_2O and CH_4 emissions during swine manure composting without forced aeration–effect of compost pile scale. *Bioresour. Technol.*, *89*, 109-114.

[74] Fukumoto, Y., Rom, H. B. & Dahl, P. (2003). Relationship between gas depth profiles in compost heap and gas emission. *Agr. Eng. Intl.*, *5*, (EE 03 004), 1-13.

[75] Fukumoto, Y., Suzuki, K., Osada, T., Kuroda, K., Hanajima, D., Yasuda, T. & Haga, K. (2006). Reduction of nitrous oxide emission from pig manure composting by addition of nitrite-oxidizing bacteria. *Environ. Sci. Technol.*, *40*, 6787-6791.

[76] Haga, K. (1998). Animal waste problems and their solution from the technological point of view in Japan. *Jarq - Jpn. Agric. Res. Q.*, *32*, 203-210.

[77] Hao, X., Chang, C., Larney, F. J. & Travis, G. R. (2001). Greenhouse gas emissions during cattle feedlot manure composting. *J. Environ. Qual.*, *30*, 376-386.

[78] He, Y., Inamori, Y., Mizuochi, M., Kong, H., Iwami, N. & Sun, T. (2001). Nitrous oxide emissions from aerated composting of organic waste. *Environ. Sci. Technol.*, *35*, 2347-2351.

[79] Houghton, J. T., Meira Filho, L. G., Bruce, J., Lee, H., Callander, B. A., Haites, E., Harris, N. & Maskell, K. (1995). *Climate Change 1994: radiative forcing of climate change and an evaluation of the IPCC IS92 emission senarious*. Cambridge University Press, Cambridge, UK.

[80] Kester, R. A., de Boer, W. & Laanbroek, H. J. (1997). Production of NO and N_2O by pure cultures of nitrifying and denitrifying bacteria during changes in aeration. *Appl. Environ. Microbiol.*, *63*, 3872-3877.

[81] Kirchmann, H. & Widén, P. (1994). Separately collected organic household wastes - chemical-composition and composting characteristics. *Swed. J. Agric. Res.*, *24*, 3-12.

[82] Kuroda, K., Osada, T., Yonaga, M., Kanematu, A., Nitta, T., Mouri, S. & Kojima, T. (1996). Emissions of malodorous compounds and greenhouse gases from composting swine feces. *Bioresour. Technol.*, *56*, 265-271.

[83] Myers, R. J. K. (1975). Temperature effects on ammonification and nitrification in a tropical soil. *Soil Biol. Biochem.*, *7*, 83-86.

[84] Osada, T., Kuroda, K. & Yonaga, M. (2000). Determination of nitrous

oxide, methane, and ammonia emissions from a swine waste composting process. *J. Mater. Cycles Waste Manag.*, *2*, 51-56.

[85] Pearson, J. & Stewart, G. R. (1993). The deposition of atmospheric ammonia and its effects on plants. *New Phytol.*, *125*, 283-305.

[86] Remde, A. & Conrad, R. (1990). Production of nitric oxide in *Nitrosomonas europaea* by reduction of nitrite. *Arch. Microbiol.*, *154*, 187-191.

[87] Venterea, R., Rolston, D. E. & Cardon, Z. (2005). Effects of soil moisture, physical, and chemical characteristics on abiotic nitric oxide production. *Nutr. Cycl. Agroecosyst.*, *72*, 27-40.

INDEX

A

AA, 45
abatement, 48, 49, 174
abiotic, 209, 230
absorption, 74
Abundance, vii, 191, 199, 205
AC, 52
accountability, x, 2
accounting, 33, 35, 36, 42, 58, 180
acetate, 134, 137, 140, 142
acid, 81, 88, 100, 105, 123, 134, 155, 160, 214, 219, 226, 228
acidification, 115
Actinobacteria, 202
actinomycetes, 74
additives, 4, 9, 30
adjustment, 133
adsorption, 102
AE, 50, 64
aerobic, xi, 12, 46, 97, 122, 143, 144, 149, 226
Africa, ix, 61, 120
Ag, 79, 94, 109
agar, 197
agent, 137
agents, 80
aggregates, 8, 175, 176
aggregation, 57, 170, 171, 176, 182, 183, 185
aggregation process, 182, 183, 185
agriculture, ix, xii, 1, 17, 18, 19, 27, 28, 46, 47, 48, 49, 58, 61, 64, 81, 116, 165, 167, 186, 192, 215
aid, 140
air, x, 2, 5, 17, 27, 29, 45, 74, 100, 105, 106, 166, 167, 228
air quality, x, 2, 17
AJ, 46, 52
AL, 22, 51, 64
Alabama, 60, 148
Alberta, 52, 55
alcohol, 123
alfalfa, 89, 214
algae, 9, 43, 77
alkaline, 133, 219
alkaline phosphatase, 133
alkaloids, 84
alternative, x, 2, 40, 44, 49, 51, 54, 60, 67, 79, 132, 141
alternatives, 45
aluminum, 129, 138

amendments, x, 11, 72, 75, 77, 79, 82, 83, 89, 92, 95, 102, 107, 109, 110, 114, 116, 118, 119, 120, 129, 153
amino, 214
amino acid, 214
ammonia, xiii, 48, 52, 68, 88, 92, 100, 103, 105, 106, 167, 168, 217, 219, 220, 223, 227, 228, 230
ammonium, x, 72, 85, 106, 138, 195, 219
AMS, 57, 59
AN, 63, 64
anaerobe, 214
anaerobic, 7, 11, 12, 40, 48, 55, 59, 98, 139, 209, 215
anaerobic bacteria, 11
Animal feed, 8, 15
animal feeding operations, 25, 31, 49, 58, 60
animal health, 210
animal husbandry, 48, 192, 209
animal tissues, 73
animal waste, 47, 51, 55, 218, 219, 221
animals, ix, 1, 3, 12, 31, 41, 73, 81, 97, 100, 108, 131, 166, 192, 214
annealing, 196
annual rate, 148
antagonistic, 80, 84, 93
anthropogenic, 219
antibiotic, xii, xiii, 111, 191, 192, 193, 194, 198, 199, 207, 208, 209, 211, 212, 213, 214, 215, 216
antibiotic resistance, xii, 191, 192, 193, 198, 199, 207, 208, 209, 211, 212, 213, 214, 215, 216
antibiotics, 91, 102, 107, 108, 192, 193, 194, 207, 209, 214, 215
appendix, 175
aquaculture, 10
aqueous solution, 155
aqueous solutions, 155
aquifers, 102
ARB, 195, 214
arid, 50
Arizona, 63
Arkansas, 11
ARS, 12, 121
arthropods, 74, 77
ash, 91, 129, 154
Asia, 50, 81
Asian, 10, 109, 119
Asian countries, 10
assessment, 45, 47, 55, 61, 159, 173
assumptions, 24, 169, 173
atmosphere, 27, 78, 103, 179
AU, 5
Austria, 69, 120
authority, 58
availability, 20, 21, 24, 34, 44, 51, 85, 87, 93, 96, 102, 123, 129, 140, 155, 161
awareness, 19, 28

B

bacterial, xi, xii, xiii, 72, 74, 80, 94, 96, 101, 111, 191, 192, 193, 194, 198, 200, 202, 203, 206, 208, 211, 212, 213, 215, 216
banks, 26, 39, 43
barley, 90, 115
Bayesian, 197
beef, 15, 38, 49, 50, 55
behavior, 49
Beijing, 158
Belgium, 196
benchmark, 175
benchmarking, xii, 166
benefits, 3, 4, 5, 6, 14, 17, 31, 34, 37, 39, 43, 59, 84, 108
binding, 214
bioavailability, 41, 102, 118, 123, 130, 143
biocontrol, 96
biodegradable, 10

biodiesel, 9
biodiversity, 112
biogas, 15, 47
Biogas, 15, 66
biogeography, 111
biological activity, 18
biological control, 109, 117
biological processes, 29
biomass, ix, 1, 3, 44, 74, 87, 94
biosphere, 73
biota, x, 72, 73, 77, 101
biotic, 209
biotic factor, 209
birds, 81
blot, 206
Boron, 6
bottom-up, xii, 166
Brazilian, 112
breakdown, 115, 117
breeding, 166
British Thermal Units (BTUs), 16
BTUs, 16
buffer, 37, 38, 134, 137, 140, 142, 194, 195
buildings, 116
bulbs, 120
Burkina Faso, 50
buyer, 29
by-products, 83, 94, 173

C

calcium, 138
Canada, 55, 115, 163
canals, 78
candidates, 96
Cape Town, 120
carbohydrates, 74
carbon, 9, 77, 82, 86, 89, 100, 105, 107, 228
Carbon, 9, 16, 77, 113
carbon dioxide, 100, 105, 228
carbon molecule, 77
carbonates, 124
carrier, 17
case study, xi, 122
cassettes, 207, 211
cast, 47
cation, 8
cats, 11
CE, 50
CEC, 8
cell, 223
cellulosic, 94
Census, 21, 65
Central Europe, 113
CH4, 229
channels, 78, 95
chemical bonds, 133
chemical properties, 116
chemicals, 36, 79, 102, 113
chicken, 56, 88, 89, 92, 105, 115
China, 94, 115, 158
chitin, 94
Chloride, 6
chrysanthemum, 84, 113
CL, 52
classes, 130, 169, 172, 179, 180, 181, 183
classification, 130, 169, 171, 172, 179, 180, 185
clay, 8, 51, 57
cleaning, 105
climate change, 228, 229
clone, 202, 208, 210
cloning, 208
cluster analysis, 171, 172, 177, 185, 200, 201
clustering, 171, 172, 179, 197
clusters, xii, 165, 169, 171, 172, 178, 179, 180, 181, 182, 183, 185
CO2, 78, 219
coal, 129
cocoon, 87
codes, 167

Columbia, 50
combustion, 129
communication, 65
communities, xiii, 10, 13, 76, 80, 92, 119, 120, 192, 193, 208, 211, 212, 213, 216, 222
community, xii, xiii, 91, 94, 95, 115, 116, 191, 192, 193, 196, 202, 207, 215, 216, 220
competition, 166, 220
complex interactions, 73, 82
compliance, 69
components, 24, 32, 73, 117, 219, 225
composites, 11
composition, 23, 63, 73, 95, 123, 132, 148, 155, 160, 161, 167, 193, 202, 206, 208, 222, 229
compost, ix, xiii, 1, 3, 12, 13, 40, 48, 77, 80, 90, 93, 96, 101, 103, 114, 140, 150, 162, 207, 218, 222, 223, 225, 226, 227, 228, 229
composting, ix, xiii, 40, 54, 55, 83, 96, 97, 105, 107, 111, 114, 115, 118, 217, 219, 220, 221, 222, 223, 224, 225, 226, 227, 228, 229, 230
compounds, xiii, 9, 73, 84, 91, 95, 96, 98, 99, 102, 130, 133, 137, 138, 139, 141, 155, 157, 158, 160, 162, 217, 229
concentration, xiii, 46, 57, 96, 100, 105, 129, 137, 150, 166, 205, 218, 220, 221, 224
conditioning, 82
confidence, 24, 197
confinement, 16, 22, 28, 31, 32, 41, 44
Congress, iv, 50, 66, 111
conservation, 20, 28, 36, 37, 58
consumers, 12
contaminant, 103
contaminants, 32, 54
contamination, 100, 102, 104, 125
control, xiii, 10, 12, 16, 30, 31, 37, 38, 41, 50, 51, 57, 69, 84, 92, 93, 97, 107, 109, 110, 111, 113, 144, 148, 192, 218, 222, 224
conversion, 15, 16, 106, 168
conversion rate, 16
copper, 87
corn, 38, 70, 87, 104, 110, 119, 163
correlation, xiii, 94, 136, 137, 148, 177, 196, 218, 220, 221
correlation coefficient, 136, 137
Corynebacterium, 203, 204
cost-effective, 42, 43, 102
costs, x, 2, 18, 27, 31, 32, 34, 37, 38, 39, 40, 41, 42, 55, 56, 58, 59, 61, 107, 166, 168, 178, 184
costs of production, 58
countermeasures, 219, 227
covering, 73, 105, 203
cows, 11, 15, 30, 39, 69, 123
CR, 196
crab, 82
CRC, 112
credit, 42
crop production, ix, 1, 3, 9, 14, 102, 107, 122, 146, 180, 182, 183, 184, 190
crop residues, 93
crops, 12, 22, 38, 52, 80, 83, 84, 90, 99, 104, 107, 108, 110, 129, 177, 178, 179
cross-validation, 172, 180, 181
Crotalaria, 119
CRS, 188
crustaceans, 82
crystalline, 124, 139
cultivation, 10, 16, 67, 159, 206, 208, 209
culture, 81, 226
current limit, 107
customers, 11
cycles, 194, 196
cycling, xi, xiii, 61, 72, 91, 92, 192
cycloheximide, 194, 197
cysts, 91

D

dairies, 23, 32
dairy sector, 53
Dallas, 48
data set, 168, 171, 172, 177, 179
database, 197
decisions, 62, 140
decomposition, 74, 84, 85, 91, 92, 95, 98, 99, 115
deconvolution, 141
deficiency, 102
deficit, 43
definition, 169, 178
deforestation, 73
degradation, xi, 74, 82, 96, 99, 102, 122, 123, 148, 161, 166, 170, 219
degrading, 83, 167
denaturation, 196
denitrification, 98, 106
denitrifying, 229
density, xi, 68, 72, 89, 94
Department of Agriculture, 45, 51, 54, 56, 60, 62, 65, 66
deposition, 26, 29, 39, 43, 44, 230
deposition rate, 26
depressed, 120
desorption, 102, 145, 155
detection, 179, 194, 211, 216
developed countries, x, 2
developed nations, x, 2
developing nations, x, 2
diarrhea, 192
diet, 12, 52, 61
dietary, 30, 31, 57, 128
diets, 30, 41, 56, 129, 131, 139, 155, 160, 162, 163
diffraction, 139
digestion, 7, 40, 48, 59, 91, 99, 135, 139
discriminant analysis, 180
discrimination, 195
diseases, 12
distribution, 20, 106, 127, 128, 135, 144, 152, 154, 156, 180
diversification, xii, 166, 175, 177, 183
diversity, 77, 134, 169, 170, 179, 185, 193, 208, 211, 212, 213, 216
DNA, xii, xiii, 120, 130, 132, 133, 135, 142, 146, 192, 193, 194, 195, 196, 200, 202, 203, 204, 205, 207, 208, 212
dominance, 139
dosage, 211
drainage, x, 72, 75, 84, 98, 103, 129
drinking, 17, 167
drinking water, 18, 167
drought, 8, 79
dry matter, 105, 127, 128, 228
drying, 105, 154
dung, 93, 107, 132, 161
duration, 117, 220, 224

E

E. coli, 194, 195, 196, 204, 206, 209
earth, 73
earthworm, 81, 87, 91, 110
earthworms, 74, 77, 82, 88, 91, 110, 112, 120
eating, 81
Ecological Economics, 186
ecology, 111
economic performance, xii, 166, 168
Economic Research Service, 45, 54, 56, 60, 62
economics, 46, 49, 53, 54, 55, 68
ecosystem, x, 2, 5, 8, 112
ecosystems, 46, 82, 94, 154, 167
effluent, 9, 61, 87
effluents, 167
egg, 20
electric utilities, 186
electricity, 15, 16
electron microscopy, 114

elephants, 12
emerging economies, x, 2
emission, ix, xiii, 53, 100, 105, 106, 115, 116, 167, 168, 173, 217, 219, 220, 221, 222, 223, 224, 225, 226, 227, 229
encoding, 119, 193, 207, 213, 216
energy, x, 2, 7, 13, 14, 15, 16, 40, 43, 46, 47, 74, 78, 81, 89, 97, 105, 107, 140
England, 58
enteritis, 209
environment, ix, xii, xiii, 1, 38, 59, 61, 78, 79, 82, 96, 97, 99, 103, 105, 107, 122, 144, 166, 167, 168, 192, 213, 214
environmental conditions, 135
environmental degradation, 166
environmental effects, xi, 37, 60, 121
environmental impact, iv, ix, 31, 51, 59, 60, 64, 115, 129, 153, 166
environmental issues, 22, 25, 167
Environmental Protection Agency, 66, 67
environmental regulations, 58
environmental standards, 68
enzymatic, xi, 122, 132, 133, 134, 136, 137, 143, 156, 157, 158
enzymes, 30, 133, 134, 135, 137
EPA, 6, 31, 66, 67
EQIP, 62
equilibrium, 100
ER, 51
erosion, 19, 29, 34, 35, 38, 41, 42, 73, 99, 104, 153
Escherichia coli, 101, 212, 213, 214, 215
ESO, 65
essential oils, 116
estimating, 163, 169, 172
estuarine, 213
ET, 47, 52, 62
EU, 167
Europe, 68
European Commission, 165
European Union, 167
eutrophication, 18, 33, 100, 146, 153, 167
evaporation, 8
evolution, 181, 183, 184, 211, 215
exclusion, 38, 121
excretion, 30, 31, 41, 52, 56, 57, 62, 74, 128, 160, 211, 218
externalities, 27
extraction, 9, 123, 124, 125, 126, 129, 130, 132, 146, 154, 159, 169

F

FA, 58
faecal, 116
family, 94, 178, 195, 203, 208
FAO, 228
farm size, 19
farmers, 3, 27, 31, 36, 37, 64, 81, 84, 140, 166, 173, 218, 221, 222
farming, ix, xii, 1, 4, 5, 12, 37, 50, 51, 52, 61, 113, 115, 166, 168, 169, 170, 171, 173, 175, 176, 177, 180, 182, 183, 184, 185
farmland, 36, 215, 218
fats, 74
fatty acids, 91, 95, 96, 109
fauna, 10, 73, 78, 113
February, 63, 113
feces, ix, 1, 3, 128, 138, 155, 161, 206, 208, 211, 218, 222, 229
feed additives, 4, 9, 30
feeding, xi, 26, 39, 62, 63, 72, 79, 94, 96, 101, 110, 111, 128, 168, 193
feedstock, x, 2, 3, 40, 43
fermentation, 12, 82, 223
fertility, ix, 77, 80, 91, 103, 104, 192
fertilization, 100, 111, 113, 192, 193, 211
fertilizer, ix, 7, 9, 14, 17, 19, 21, 24, 25, 26, 34, 35, 36, 44, 45, 62, 84, 89, 94,

96, 97, 99, 106, 107, 115, 142, 144, 146, 151, 157, 209, 218, 219, 224, 227
fertilizers, 3, 4, 7, 8, 26, 84, 88, 102, 103, 107, 118, 140, 167, 178
fiber, 11
film, 78
fines, 173
fingerprinting, xiii, 192
fires, 186
firms, 169, 173, 174, 186, 187
fish, 10, 81, 115
FL, 51, 69, 112
flexibility, x, 2, 36
flora, 10, 92, 193
flora and fauna, 10
flow, 37, 98, 104
fluctuations, 144
flue gas, 129
fluorogenic, 216
food, ix, 2, 49, 74, 77, 82, 89, 92, 95, 96, 112, 210, 212, 214
forage crops, 140
Forestry, 114, 218, 228
forests, 212
formaldehyde, 91
formamide, 195
fossil, 97, 107, 186
fossil fuel, 97, 107
Fourier, 158
fractional composition, 40
fractionation, xi, 122, 123, 124, 125, 126, 128, 132, 134, 137, 143, 148, 154, 155, 156, 157, 158, 161, 162, 163
France, 165, 170, 177, 179, 184, 185, 194
freshwater, 18, 50
fruits, 101, 112
frustration, 25
FT-IR, 157
fuel, 97, 107, 186
fumigation, 111
funding, 37, 41
fungal, xi, 72, 80, 89, 137
fungi, xi, 72, 74, 77, 78, 89, 90, 92, 112, 114, 115, 116, 117, 119
fungus, 76, 114, 116

G

garbage, 83, 90, 93
gas, 73, 129, 229
gases, 100, 103, 108, 226
gasification, 7
GC, 196, 208
gel, 194, 196, 200, 203, 213
gels, 195, 196, 200
GenBank, 194, 197, 203
gene, xii, xiii, 192, 194, 195, 196, 198, 199, 200, 201, 202, 203, 205, 206, 208, 211, 212, 214, 215, 216
gene amplification, 196
gene transfer, 212
generalization, 180
generation, xiii, 7, 9, 13, 14, 15, 23, 40, 41, 43, 217, 220
genes, xii, xiii, 192, 193, 194, 195, 198, 199, 200, 202, 204, 205, 207, 208, 209, 210, 211, 213, 214, 215, 216
genomic, 195
Georgia, 61, 117
Germany, xii, 160, 191, 193, 194, 195, 196, 209
germination, 91
GHG, 105
Gibbs, 101, 112
GIS, 62
GL, 53, 61, 67
global warming, xiv, 218, 219
glycosides, 84
goals, 44, 136
government, iv, 28, 38, 44
grains, 137, 160
Gram-negative, 206, 209, 215

Gram-positive, 209
grass, 39, 97, 104, 119, 145
grasses, 38
grassland, 110, 111, 162
grasslands, 120
grazing, 26, 29, 38, 39, 41, 43, 44, 53, 132, 161
Great Britain, 12
greenhouse, 9, 43, 66, 100, 167, 219, 226, 229
Greenhouse, 66, 67, 219, 229
greenhouse gas, 9, 43, 66, 167, 219, 226, 229
greenhouse gases, 229
ground water, 98, 100, 103, 108, 210
groundwater, 142, 167, 211
grouping, 171
groups, 74, 77, 111, 127, 147, 150, 168, 171, 172, 173, 176, 177, 179, 180, 220
growth, x, xiii, 3, 8, 10, 18, 46, 48, 72, 74, 86, 88, 90, 92, 97, 103, 108, 111, 118, 119, 145, 209, 217, 219, 220, 221, 222, 226, 227
growth rate, 226
guidance, 65
Gulf of Mexico, 19, 47
gut, 111, 192, 214

H

HA, 52
habitat, 77
handling, iv, x, 2, 14, 20, 21, 22, 23, 29, 40, 42, 65, 66, 85, 97, 98, 108, 110
harbour, 101
harm, 167
harmful effects, 98
harvest, 77, 79, 91, 101
harvesting, 101
hazards, 79, 219
health, ix, x, 1, 5, 8, 12, 17, 43, 49, 51, 69, 72, 73, 80, 83, 100, 103, 108, 210
heat, 14
heavy metal, 58, 102, 118
heavy metals, 58, 102, 118
hemp, 84
herbivores, 77, 95
heterogeneity, 170, 171, 176, 182, 185
heterogeneous, 169, 183
heterotrophic, 220
heterotrophic microorganisms, 220
high temperature, 222
histogram, 177
hog, 31, 52, 62, 96, 115
hogs, 31, 52, 123
holistic, x, 2, 3, 13, 43, 44
homogenous, 169, 181, 185
horizon, 219
hormones, 55
horse, 10, 11, 107
host, 79, 94, 117, 206, 207
House, 99
household, 228, 229
household waste, 228, 229
housing, 9, 39, 167
human, ix, xiii, 1, 16, 17, 18, 43, 51, 101, 103, 108, 192, 207, 210, 214
human activity, 18
humans, 100, 108, 210
humus, 74
husbandry, 48, 69, 168, 192, 209, 218
hybridization, 206
hydrogen, 95, 100, 117, 118, 139, 141
hydrogen sulfide, 95, 117, 118
hydrologic, 45
hydrolysis, xi, 122, 124, 133, 134, 135, 136, 137, 138, 141, 143, 156, 157, 158, 160
hydrolyzed, 124, 133
hydroxide, 138
hydroxides, 138
hydroxylapatite, 138
hygienic, 101, 108

hypothesis, 170, 180
hypoxia, 19
Hypoxia, 19

I

id, 50
identification, 206
identity, 200, 204
Illinois, 52
imbalances, 21
immobilization, 100, 147
implementation, 27, 32, 65, 66, 165, 173, 177, 179, 185
imports, 140
in situ, 214
in vitro, 214
incentive, 166
incidence, 39, 93, 120
incubation, xi, 122, 135, 137, 143, 144, 147, 149, 194, 206
incubation period, 144
incubation time, 143
India, 71, 118
Indian, 113, 117
Indiana, 46
indication, 25, 208
indicators, 95, 113, 226
indigenous, xiii, 214, 217, 220, 221, 222, 227
industrial, 94
industry, 4, 219
ineffectiveness, 25
inefficiency, 181, 185
inertia, 179
infections, 101, 108
infectious, 192
infectious disease, 192
infectious diseases, 192
infrared, 158
ingestion, 91
inhibitors, 221
injection, 35, 42, 168
injury, iv
innovation, 11
inorganic, x, 24, 26, 35, 72, 77, 85, 86, 98, 103, 106, 107, 119, 123, 124, 126, 127, 133, 137, 138, 141, 144, 146, 147, 155, 157, 158, 159, 222, 223
inositol, 130
insects, 12
insertion, 212
insight, 21, 139, 208
interaction, 77, 147
interactions, x, 2, 73, 80, 82
interrelations, 28
intrinsic, 212
intuition, 34
inventories, 13, 15, 16
inversion, 37
invertebrates, 73, 82
ionic, 130
ions, 130
IPCC, 229
irradiation, 92
irrigation, 98
isolation, 206, 216
isotope, 129
isotropic, 141
Italy, 228
ITO, 120

J

Japan, xiii, 196, 217, 218, 219, 228, 229
Jordan, 48, 117
JT, 51, 56, 63, 64
judge, 168

K

kelp, 84
killing, 96
King, 212

KL, 47

L

L2, 139, 160
labour, xii, 166, 169, 178, 185
lactating, 30
lactation, 30
Lactobacillus, 202, 208
Lafayette, 46
lagoon, 28
lakes, 18
land, xi, xii, 3, 11, 18, 19, 20, 21, 23, 24, 33, 37, 38, 42, 43, 49, 51, 53, 56, 57, 62, 65, 66, 68, 101, 104, 110, 111, 121, 123, 166, 167, 168, 169, 177, 179, 185
land use, 57, 62, 169, 185
large-scale, 34, 57
larvae, 113
larval, 112
leachate, 162
leaching, 7, 17, 98, 100, 103, 106, 111, 147, 152, 168
leakage, 77
legislation, 57
legume, 3, 35, 84, 104
legumes, 104
limitations, 79, 107, 167
linear, 169, 175, 181, 188
linear programming, 169, 175, 188
linkage, 216
lipid, 123
lipids, 84
Listeria monocytogenes, 101
litigation, 65
Livestock, xiii, 9, 16, 41, 45, 49, 53, 54, 57, 58, 59, 61, 178, 217, 228
LM, 50, 54, 57, 59, 60
local government, 28
localization, 206
location, 59, 62, 104, 108, 214
locus, 216
logistics, 20
London, 47, 51, 64, 116
long period, 24
long-term impact, xi, 122
losses, xiv, 7, 13, 14, 17, 18, 19, 21, 22, 24, 25, 27, 31, 33, 35, 36, 38, 39, 41, 42, 44, 50, 54, 55, 58, 79, 98, 100, 106, 111, 129, 218, 219, 224, 225
low temperatures, 98

M

M1, 204
machinery, 37
macronutrients, 87
Madison, 196
magnesium, 138
magnetic, iv, xi, 122, 153, 154, 157, 158, 159, 161, 162
Maine, 121, 143
maintenance, x, 2, 30, 37, 39
maize, 57, 95, 115
management practices, x, xi, 2, 24, 25, 32, 36, 38, 44, 47, 99, 108, 122
manganese, 87
Manganese, 6
Manhattan, 48
manipulation, 128
manufacturer, 194
manufacturing, 40
marine environment, 50
market, x, 2, 57, 166, 178
marketing, 44, 57
markets, 40, 43
matrix, 96, 124, 171, 196
maturation, xiii, 217
measurement, 116, 134, 159, 169, 171, 173, 174, 175, 176, 181, 182, 183, 184, 185, 187, 190
measures, 21, 103, 167, 169, 170, 175, 192

meat, 19, 166, 168, 173, 193, 197, 199, 207, 209
media, 27, 58, 59, 116, 206
melt, 36
Merck, 194
metabolic, 77
metabolite, 205
metabolites, 205
metals, 87, 96, 102
methane, 9, 14, 27, 43, 95, 99, 167, 230
methanol, 7
microbes, 82, 113
microbial, 47, 50, 62, 74, 82, 88, 93, 94, 98, 100, 101, 108, 116, 193, 208, 214, 215, 216, 219
microbial communities, 193
microbial community, 215
microcosm, 110
microflora, 192
micronutrients, 3, 7, 87, 92
microorganisms, xiii, 3, 8, 13, 17, 73, 82, 90, 133, 217, 219, 220
micro-organisms, 92
migration, 27
milk, 19, 20, 30, 39, 57, 178, 214
millet, 119
mineralization, xi, 24, 72, 73, 78, 82, 86, 92, 98, 143
mineralized, x, 72, 77, 85, 98, 108
minerals, 73, 81, 139, 145, 146
Ministry of Education, 228
missions, 29, 48, 52, 68, 105, 118, 181
Missouri, 50
mites, xi, 72, 74, 77, 88, 109, 113, 116
ML, 46, 68
MMW, 55
mobility, 102, 196, 208
modeling, 45, 50, 54
models, 20, 169, 186, 187
moisture, 230
molecular biology, 212
molecular weight, 95
molecules, 77, 147, 194, 198
molybdenum, 135
mood, 63
mortality, 32
movement, 25, 26, 37, 39, 74, 78, 96, 112
MS, 52, 55, 63
MSW, 103
municipal solid waste, 118
Musca domestica, 112
Muslim, 71
myo-inositol, 130

N

N-acety, 205
nation, 3, 21, 219
National Research Council, 58
National Science and Technology Council, 47
native population, 96
natural, 8, 10, 13, 16, 89, 99, 154, 158, 161, 168
natural enemies, 89
natural environment, 99
natural resources, 16
Natural Resources Conservation Service, 54, 65, 66
NC, 22, 68, 118
ND, 51, 130, 131
Nebraska, 56
neem, 83, 84, 109
nematicides, 79, 110, 119
nematode, 76, 79, 80, 89, 90, 91, 92, 95, 96, 101, 107, 109, 110, 111, 112, 113, 114, 115, 116, 117, 119, 120
nematodes, xi, 72, 74, 77, 78, 79, 83, 84, 88, 89, 90, 91, 92, 94, 96, 97, 109, 110, 111, 112, 113, 114, 117, 118, 119, 120
nested PCR, 203
Netherlands, 48, 54, 116
New England, 121

New York, iii, v, 111, 116, 228
Newton, vii, 53, 67, 191
nitrate, xiii, 17, 58, 100, 106, 167, 217, 219, 221
nitrates, 167, 220
nitric acid, 226
nitric oxide, 219, 225, 230
nitric oxide (NO), 225
nitrification, xiii, 97, 217, 219, 220, 221, 222, 223, 228, 229
nitrifying bacteria, 220, 222, 223
Nitrite, vii, 217, 224
nitrogen compounds, 222, 223
nitrogen fixation, 3
nitrogen-fixing bacteria, 74
nitrous oxide, xiii, 9, 99, 103, 167, 217, 219, 228, 229, 230
NMR, xi, 122, 129, 130, 131, 132, 133, 135, 136, 137, 138, 139, 140, 141, 142, 145, 146, 149, 150, 158, 159, 160, 161
NO, xiii, 217, 219, 220, 221, 222, 223, 224, 225, 226, 227, 229
nonparametric, xii, 166, 169, 172, 173, 185
normal, 18, 222
North America, 61, 111
North Carolina, 69, 118
Northeast, 94, 115
NPP, 30
NRC, 23, 58
nuclear, xi, 122, 153, 154, 157, 158, 159, 161, 162
nuclear magnetic resonance, xi, 122, 153, 154, 157, 158, 159, 161, 162
nuclease, 133, 134, 194
nucleotide sequence, 197
nucleotides, 195
nutrient cycling, xi, 61, 72, 91, 92
nutrient flow, 67
nutrient imbalance, 21
nutrient transfer, 88
nutrition, 35, 48, 63, 67

O

OB, xiii, 217, 220, 221, 222, 223, 227
observations, 136, 141, 148, 149, 169, 171, 172, 179, 188
odors, 63
OECD, 167, 187
Ohio, 65
oil, 8, 73, 77, 83
oils, 116
Oklahoma, 163
online, 109
OR, 188
organic chemicals, 113
organic compounds, 73, 82
organic matter, ix, 3, 7, 9, 50, 73, 77, 79, 80, 82, 83, 88, 92, 99, 108, 113, 115, 117, 158, 160, 219
Orthophosphate, 130, 141
orthophosphates, 139
outliers, 177
oxidation, 219
oxidative, 135
oxide, xiii, 99, 103, 167, 217, 219, 228, 229, 230
oxides, 124, 147, 228
oxygen, 19, 97, 99, 219, 220
ozone, 219, 226

P

PA, 62, 121
Pacific, 118
packaging, 4
paramagnetic, 130
parameter, 152
parasites, 79, 80, 117
Parasites, 118
Paris, 187
particles, 18, 74, 78
partition, 179

pasture, xi, 25, 26, 33, 38, 39, 43, 44, 48, 60, 122, 132, 148, 153, 158
pastures, 25, 29, 39, 43, 44, 149
pathogenic, 101, 112, 209
pathogens, x, xiii, 4, 9, 17, 18, 22, 32, 42, 72, 75, 80, 107, 167, 192
PCA, 177
PCR, xii, xiii, 192, 193, 194, 195, 196, 202, 203, 206, 207, 208, 211, 216
penicillin, 214
Pennsylvania, 62, 121
permit, 69
pest control, 12
pest populations, 80
PG, 54, 59
pH, 8, 105, 109, 128, 130, 133, 134, 137, 140, 142, 144, 219
pharmaceutical, 213
phosphatases, 133, 134
phosphate, 14, 69, 86, 100, 129, 138, 141, 146, 158, 159, 161
Phosphate, 48, 61, 65
phosphates, 139, 156, 157, 158
phosphonates, 130, 131
phosphorous, ix, 52, 92
photochemical, 226
phylogenetic, 204
phylum, 200, 202
physical properties, 84
physicochemical, 143
phytoremediation, 102
pigs, xii, 62, 131, 168, 185, 191, 205, 213
PL, 145, 148, 152, 153
planning, 65
plants, ix, x, 1, 3, 13, 18, 72, 74, 77, 78, 83, 85, 86, 88, 90, 92, 101, 106, 107, 123, 133, 140, 160, 230
plasma, 135
plasmid, 212, 213, 215
plasmids, 192, 206, 207, 211, 212, 216
play, ix, 1, 6, 75, 82, 220
PM, 130
Poland, 113, 115
policymakers, 25
pollutant, 17, 27
pollutants, 18, 26, 170
pollution, 5, 17, 25, 29, 33, 39, 49, 57, 63, 67, 68, 73, 166, 167, 168, 174, 175, 182, 183, 228
polymerase, 195
polynucleotide, 134, 135
Polyphosphate, 130, 141, 145
pools, 124, 148, 159
poor, 98, 103, 151
population, xi, 72, 79, 87, 89, 91, 93, 96, 208, 212, 215
population density, 91
pore, 8, 78
pores, 78
pork, 46, 56, 61
Portugal, 191
Potash, 48, 61
potassium, xi, 3, 4, 7, 14, 72, 86, 98, 104
potato, 90, 112, 114, 118, 134
potatoes, 79, 97
powder, 139
power, 40, 41
PPI, 14, 20, 21, 22, 61
pragmatic, 103
precedents, 57
precipitation, x, 72, 86, 100, 102
predators, 74, 80, 92
prediction, 153
pressure, 105, 166, 216
prevention, 36, 222, 227
preventive, 173
price effect, 68
prices, 14, 166, 169, 184
principal component analysis, 171
PRISM, 194
pristine, 207
probability, 172
probe, 194
producers, 24, 25, 28, 31, 33, 34, 38, 39, 44, 63, 168, 173, 184
production costs, 166, 184

production technology, xii, 166, 169, 173, 174, 182, 185, 188
productivity, 113, 166, 186
profit, 39
profitability, 46, 63
profits, 39
program, 51, 69, 194
programming, 187
proliferation, xiii, 218, 222
property, iv, 27, 29, 52, 60, 62, 117
prophylactic, 207
protection, 46, 48, 63, 64, 167
protein, 8, 12, 52, 81, 140
proteins, 74, 133, 193, 214
Proteobacteria, 202
protocols, 194
Pseudomonas, 93, 101, 210, 212, 214
Pseudomonas spp, 101, 210
PT, 59
public, 19, 166, 219
public awareness, 19
purification, 18, 219
pyrophosphate, 130, 132, 134, 141, 145

Q

quality improvement, 47
quality of life, 69

R

RA, 46, 49
rain, 26, 35, 98, 100, 219, 226
rainfall, 23, 55, 100, 162
range, 4, 12, 17, 40, 58, 79, 130, 134, 139, 147, 185, 198, 207, 208, 220
rangeland, 61
rape, 159
RB, 60
RC, 53, 58
RDP, 197, 200, 202, 203, 204
reading, 210
reagent, 221
reagents, 221
real time, xii, 192, 193
reclamation, 10, 16, 67
recombination, 216
recovery, 106, 132
recreational, 167
recycling, xi, 74, 103, 121, 123
regional, 21, 34, 166
regression, 169
regulation, 46, 53, 168, 170, 173, 184
regulations, 45, 51, 53, 167, 185
regulators, 92
relationship, 68, 73, 148, 207
relationships, 169
relevance, 6, 180
renewable energy, ix, 1, 3, 7
repair, 39, 170
reproduction, 112, 118
Research and Development, 228
reservoir, 209, 211
residential, 52, 60, 62
residuals, 103
residues, 77, 88, 90, 93, 115, 119, 123, 125, 140, 141
resin, 124
resistance, vii, xii, xiii, 80, 191, 192, 193, 194, 197, 198, 199, 205, 207, 208, 209, 210, 211, 212, 213, 214, 215, 216
resolution, 151, 196
resources, x, 2, 4, 16, 23, 25, 32, 78, 96, 174
respiration, 78
retention, xi, 73, 84
returns, 36, 47, 103, 108, 188
returns to scale, 188
revenue, 178, 182, 183
RF, 65
rhizosphere, 78, 94, 159, 212
rice, 90, 107, 112, 117
rice field, 117
risk, xiv, 46, 98, 101, 108, 218, 227

risks, 102
rivers, 26
Rome, 228
rotations, 119
Royal Society, 47
ruminant, 41, 63
runoff, xii, 7, 17, 26, 28, 32, 35, 37, 42, 43, 46, 57, 63, 98, 102, 104, 111, 122, 125, 129, 142, 147, 153, 159, 160, 163
rural, 52, 62
Russia, 11
rye, 145
rye grass, 145

S

SA, 54, 55
sales, 20
saline, 98, 104, 194
salinity, 51
Salmonella, 101, 193, 209, 210, 212
salts, 98, 103
sample, 8, 85, 123, 125, 132, 135, 137, 142, 172, 175, 177, 179, 180
sample variance, 178
SAS, 188
sawdust, 77, 83, 94
scanning electron microscopy, 114
scheduling, 36
scores, 183, 184
SD, 140, 141, 142
search, 200, 203
searches, 195, 197
searching, 169
seaweed, 84
sediment, 17, 35, 36, 37, 38, 41, 42, 50, 55, 215
sediments, 65, 78, 112, 207, 212
seed, 172
seedlings, 159
seeds, 4, 9, 42, 79, 83, 172
selecting, 104, 177
semi-arid, 51, 61
semi-natural, 120
Senegal, 112, 119
SEPA, 7, 17
separation, 172, 213
sequencing, xiii, 192, 196, 200, 202, 208
series, 2, 171, 222, 224, 226, 227
services, iv
sesame, 84
sewage, 48, 81, 82, 91, 94, 96, 101, 102, 118
sharing, 169, 176
sheep, 11, 61, 107, 116, 123, 125, 131, 138, 161
Sheep, 5
Shigella, 204
short-term, 143, 144
Short-term, 110, 211
shrimp, 82
silver, 195
similarity, 196, 200, 202, 203
simulations, 34
sites, 12, 115, 207
sludge, 82, 91, 94, 96, 102, 118
socioeconomic, 49
sodium, 88, 134, 137, 139, 142
software, 195, 214
soil erosion, 3, 8, 36, 37, 42, 98
soil particles, 18, 78
soils, x, xi, 3, 8, 18, 22, 24, 33, 34, 35, 56, 58, 67, 72, 74, 78, 86, 87, 88, 92, 93, 94, 98, 99, 102, 104, 110, 112, 122, 123, 126, 143, 144, 146, 147, 148, 150, 152, 153, 154, 155, 160, 161, 162, 163, 193, 209
solid phase, 87
solid state, xi, 122, 138, 139, 140, 141, 157, 158
solid waste, 118
solid-state, 138, 145, 159
solubility, 56, 75, 124, 127, 132, 137, 140, 160, 162

sorption, x, 72, 86, 96, 145, 155
South Africa, 120
Southeast Asia, 50, 81
Southern blot, 206
soybean, 38, 115
soybeans, 79
Spain, 111, 165
spatial, 53
specialisation, xii, 166, 169, 182
specialization, 185
speciation, 138, 139, 153, 155, 156, 159, 161, 162
species, 13, 78, 79, 88, 89, 93, 135, 136, 138, 140, 141, 142, 143, 144, 145, 146, 149, 157, 162, 208
specificity, 170, 194, 195, 203
spectroscopy, xi, 122, 129, 130, 131, 132, 135, 137, 138, 139, 140, 142, 146, 154, 156, 158, 159, 160, 161, 162
spectrum, 5, 17, 136
speed, 100
spills, 29
SR, 54, 56, 58
stability, 127, 172, 180, 185
stabilization, 73, 82, 96, 102
stabilize, 129
stages, 95
stakeholders, 25, 44
standard deviation, 225
standards, 20, 49, 62, 68, 167, 168
Standards, 7, 20, 45
starvation, 107
statistical analysis, 180, 181
sterile, 194
steroids, 84
storage, x, 14, 20, 21, 22, 23, 27, 29, 31, 32, 36, 42, 49, 65, 66, 68, 72, 84, 85, 98, 99, 105, 107, 127, 137, 167, 193, 205, 208, 212, 215, 216
stormwater, 57
strain, 112
strains, 112, 210, 211, 212
strategies, 48, 56, 63, 79, 110, 123, 127
stratosphere, 219
streams, 25, 26, 104
strength, 130, 141, 184
structural changes, 166
structural characteristics, xii, 166
subsidies, 48
substances, 3, 4, 7, 8, 10, 13, 17, 213
substitutes, 97
substitution, 58, 59
substrates, 88, 89, 90, 91, 92
subsurface flow, 37
success rate, 40
Sudan, 119
sugar, 94
sugarcane, 83, 93
sulfate, 129
sulfonamide, xii, 192, 207, 210, 212, 215, 216
sulfonamides, 205, 209, 215
sulphur, xi, 72, 87, 100
summer, 110
Sun, 229
supernatant, 125
superposition, 181
supplemental, 25, 26, 31, 35
supplements, x, 72, 85
supply, 4, 84, 90
suppression, 92, 119
surface area, 74
surface water, 17, 18, 23, 25, 27, 32, 39, 48, 63, 64, 98, 100, 102, 103, 104, 122, 146, 153, 167
surplus, xii, 166, 170, 173, 176, 178, 179, 182, 183, 185
survival, 73, 96, 212
Switzerland, 215
synergistic, x, 2
synergistic effect, x, 2

T

tanks, 22, 27, 193, 206
taxonomy, 206
tea, 50, 83, 93
technology, xii, 58, 59, 61, 106, 153, 166, 173, 174, 182, 183, 185
TEM, xii, 192, 193, 194, 195, 198, 199, 205, 208, 209
temperature, 83, 98, 100, 107, 114, 194, 196, 220, 222, 226, 228
temporal, 52
tetracycline, 206, 210, 211, 214, 216
Texas, 12, 39, 45, 48, 53, 54, 57, 59, 60, 61, 65, 68, 69
TF, 68
therapy, xiii, 192
thinking, 221
Third World, 111
threat, xiii, 42, 192
threatened, 167
thresholds, 34
time, x, xiii, 2, 7, 11, 12, 18, 24, 25, 33, 85, 91, 98, 99, 103, 106, 108, 124, 125, 143, 166, 171, 172, 183, 193, 194, 209, 211, 216, 218, 219, 222
timing, 32, 33, 36, 226
tissue, 74, 79
TJ, 56, 58
TM, 63
tobacco, 83, 93, 110, 111, 120
tomato, 93, 114, 117, 118
toxic, xiii, 88, 95, 102, 217
toxicity, 88, 102, 112
toxin, 212
toxins, 80, 90
trace elements, 102
training, 172
transcription, 170
transfer, 23, 27, 88, 156, 209, 212, 213, 214
transformation, 97, 102, 143, 148, 153, 163, 214
transformation processes, 97, 214
transformations, 123, 124, 155, 156
transition, 222
translocation, 162
transmission, 214
transport, 4, 21, 29, 31, 32, 48, 98, 101, 103, 123
transportation, 60, 107
trial, 12
trichloroacetic acid, 123
tubers, 79, 90
Turkey, 5
turnover, 140, 222
turtles, 11
typology, 179

U

U.S. Department of Agriculture, 45, 51, 56, 62
UK, 48, 64, 109, 118, 213, 229
uncertainty, 20, 23, 24, 29, 44, 69
unclassified, 200, 203
uniform, 106, 226
United Nations, 49
United States, 9, 12, 17, 46, 54, 60, 65, 66, 67, 214
urea, 81, 88, 195
uric acid, 81, 105
urine, ix, 1, 3, 88, 218
USDA, 6, 12, 19, 23, 28, 32, 65, 66, 121, 163
USEPA, 7, 10, 11, 17, 27, 66, 67

V

validation, 172, 181
validity, 12, 136, 180
values, 13, 20, 27, 29, 35, 52, 60, 62, 68, 130, 133, 177, 181, 183, 197, 198

vancomycin, 212
variability, 85, 170, 171, 177, 181, 182
variable costs, 168, 178
variables, 170, 171, 177, 180, 188
variance, 177, 178
variation, 179
vector, 94, 112, 174, 175, 194, 196, 197
vegetables, 79, 101, 112
vegetation, 83, 93
velocity, 37
velvet, 84
veterinary medicine, 205, 210
Vietnam, 210
virus, 94, 120
vitamin B1, 83
vitamin B12, 83
volatilization, 27, 35, 42, 100, 106

W

Wales, 58
Warsaw, 115
waste disposal, 47
waste management, 56
waste treatment, xiii, 214, 217
wastes, 45, 77, 93, 218, 219
wastewater, 28, 32, 65, 66, 102
water quality, 17, 18, 19, 26, 34, 37, 41, 44, 45, 46, 47, 48, 52, 53, 55, 56, 64, 66, 69, 167
water resources, 23, 25, 32
water supplies, 19, 102, 108
water-holding capacity, 80
watershed, 30, 34, 39, 40, 47, 50, 54, 60, 67
watersheds, x, 2, 18, 19, 26, 40
waterways, 38
web, 12, 74, 77, 95
web sites, 12
weight gain, 20
West Africa, 61
wetlands, 38
wheat, 83, 93, 133, 134, 212
wheat germ, 134
wind, 100
wine, 81, 90, 213, 222, 229
winter, 36
Wisconsin, 60, 64
WM, 45
workers, 95
working conditions, 177
WP, 52, 137, 198

X

XANES, 138, 139, 156, 160, 161, 162
X-ray absorption, 138, 153, 154
XRD, 156

Y

yield, 77, 83, 91, 93, 103, 109, 140

Z

zinc, 87
Zinc, 6, 96
Zn, 96, 141, 146